Münchner
Geowissenschaftliche
Abhandlungen

Reihe A

Geologie
und
Paläontologie

56

AF892421

Martin PICKFORD

Revision of small Bunodont Anthracotheres (Artiodactyla, Anthracotheriidae, Microbunodontinae) from Indo-Pakistan, and Reassessment of the Age of the Caribari Deposits, Garo Hills, Bengal

Münchner
Geowissenschaftliche
Abhandlungen

In der
Reihe A
erscheinen
Originalarbeiten
und Dissertationen
aus dem Gesamtgebiet
der
Geologie
und
Paläontologie.

Münchner Geowissenschaftliche Abhandlungen

Reihe A

Geologie
und
Paläontologie

56

Martin PICKFORD

Revision of small Bunodont Anthracotheres (Artiodactyla, Anthracotheriidae, Microbunodontinae) from Indo-Pakistan, and Reassessment of the Age of the Caribari Deposits, Garo Hills, Bengal

Verlag Dr. Friedrich Pfeil
Günding, t 2026
ISSN 0177-0950
ISBN 978-3-89937-313-4

Begründet und herausgegeben von Dr. Friedrich H. Pfeil, Günding

In der **Reihe A** der **Münchner Geowissenschaftlichen Abhandlungen** werden Originalarbeiten und Dissertationen aus dem Gesamtgebiet der Geologie und Paläontologie veröffentlicht.

Die Arbeiten können in deutscher, englischer oder französischer Sprache verfasst sein. Autoren, die eine Arbeit zum Druck einreichen wollen, sollten sich vorher mit dem Herausgeber zwecks Absprache von Satzspiegel, Format und Gestaltung von Textabbildungen und Tafeln in Verbindung setzen.

Der Schriftwechsel ist ausschließlich zu richten an:
Verlag Dr. Friedrich Pfeil, Hauptstraße 12B, 85232 Bergkirchen OT Günding, Germany

Für den Inhalt der Arbeiten sind die Autoren allein verantwortlich.

Bestellungen an/orders to:

Verlag Dr. Friedrich Pfeil, Hauptstraße 12B, 85232 Bergkirchen OT Günding, Germany

Tel. ++49 (0)8131 614659-0 · Fax ++49 (0)8131 614659-1 · E-Mail info@pfeil-verlag.de

Bibliographische Information der Deutschen Nationalbibliothek

Die Deutsche Nationalbibliothek verzeichnet diese Publikation
in der deutschen Nationalbibliografie;
Detaillierte bibliografische Daten sind im Internet
unter http://dnb.dnb.de abrufbar.

Alle Rechte vorbehalten
Vervielfältigung und Wiedergabe auf jegliche Weise (grafisch, elektronisch und fotomechanisch
sowie der Gebrauch von Systemen zur Datenrückgewinnung) – auch in Auszügen –
nur mit schriftlicher Genehmigung des Herausgebers oder des Verlags.

Copyright © 2026 by Verlag Dr. Friedrich Pfeil, Günding
All rights reserved

Verlag Dr. Friedrich Pfeil, Hauptstraße 12B, 85232 Bergkirchen OT Günding, Germany
Tel. +49 8131 614659-0 · Fax +49 8131 614659-1 · E-Mail info@pfeil-verlag.de
www.pfeil-verlag.de

Druckvorstufe: Verlag Dr. Friedrich Pfeil, Günding
Druck: winterwork, Borsdorf

Printed in the European Union ISSN 0177-0950 ISBN 978-3-89937-313-4

Münchner Geowiss. Abh.	(A)	**56**	1–40	27 figs., 1 tabs.	München, März 2026

Revision of small Bunodont Anthracotheres (Artiodactyla, Anthracotheriidae, Microbunodontinae) from Indo-Pakistan, and Reassessment of the Age of the Caribari Deposits, Garo Hills, Bengal

Martin Pickford

ABSTRACT

A small species of bunodont anthracothere from the basal Late Miocene of Dhok Mila, Pakistan (Nagri Formation) was named *Microbunodon milaensis* LIHOREAU, BLONDEL, BARRY & BRUNET, 2004. The hypodigm of the species included two specimens reported as being from unknown localities. Further investigation has elucidated the localities from which the two specimens came. One is from Caribari (Garo Hills, Bengal, India), the other is from Chur Lando (Bugti region, Pakistan). The catalogue number of the fossil from Caribari was reported to be NHMUK M 19442, but it is in fact NHMUK PV OR 19042. The tooth, a right m/3, is compatible in dimensions and morphology with NHMUK PV OR 19041, the lectotype right M3/ of *Anthracotherium silistrense* PENTLAND, 1828, on which basis it is inferred that the two bunodont teeth from Caribari belong to a single species. The Caribari m/3 was included among the fossils attributed by the authors to *Microbunodon milaensis*. As such, *Mb. milaensis* becomes a junior synonym of *A. silistrense*. This conclusion prompts a revision of the small bunodont anthracotheres from Indo-Pakistan, an understanding of which requires an overview of the small anthracotheres from the Eocene and Oligocene of Europe and Southeast Asia (Myanmar, Thailand). It is demonstrated that the genus *Microbunodon* DEPÉRET, 1908, does not occur therein. A new genus is erected for the small bunodont anthracotheres from Indo-Pakistan in which there are three time successive size groups interpreted to represent distinct species. The senior name for the the largest species is Gen. nov. *silistrensis*, the medium-sized species is Gen. nov. *punjabiensis* (LYDEKKER, 1877a) and the smallest species is Gen. nov. *exiguus* (FORSTER-COOPER, 1924). Caribari also yielded the type specimen (NHMUK PV OR 19040) of the small selenodont anthracothere, *Sivameryx palaeindicus* (LYDEKKER, 1877b). The presence of Gen. nov. *silistrensis* and *S. palaeindicus* at Caribari indicates that the Bengali deposits are most likely to correlate to the late Middle Miocene to basal Late Miocene (ca 12–9.5 Ma), rather than to the Early Miocene as thought for the past two centuries.

Key Words: Miocene, Indo-Pakistan, Anthracothere, Bunodont, Selenodont, taxonomy, phylogeny, biostratigraphy, biogeography

* Centre de Recherche en Paléontologie – Paris (CR2P), Muséum national d'Histoire naturelle, CNRS, Sorbonne Université, CP 38, 8 rue Buffon, 75005 Paris, France. martin.pickford@mnhn.fr

Münchner Geowiss. Abh. (A) **56**, S. 1–40, 27 figs., 1 tabs.
© 2026 Verlag Dr. Friedrich Pfeil, Günding.

Contents

Introduction

Dentognathic remains of small bunodont anthracotheres were first described from the Miocene continental deposits of Indo-Pakistan almost two centuries ago. At Caribari (Garo Hills, Bengal, sometimes spelled Karibari) the association of a bunodont species, at the time called *Anthracotherium silistrense* PENTLAND, 1828 (based on an isolated upper molar) with a selenodont form of similar dimensions, now known as *Sivameryx palaeindicus* (LYDEKKER, 1877b) (based on a maxilla with D3/ and D4/) was the root of a great deal of confusion (Forster-Cooper, 1924; Matthew, 1929; Colbert, 1935; Pickford, 1987). Recall that this species was named only four years after the naming of the genus *Anthracotherium* CUVIER, 1822 (1824).

A bunodont lower third molar of an anthracothere from Caribari presented to the NHMUK in 1842 as part of the Cautley collection, but that remained undescribed until recently (Lihoreau et al. 2004), throws a great deal of light on the taxonomy of the bunodont species, but when described by the latter authors its locality was said to be unknown, a conclusion undoubtedly due to the fact that its catalogue number was misread as NHMUK M 19442 instead of NHMUK PV OR 19042 (see below). Examination of the OR catalogue at the NHMUK indicates that the specimen was collected at Caribari, the same locality as two other anthracothere fossils (OR 19040, a maxilla of a juvenile selenodont anthracothere, and PV OR 19041, an isolated M3/ of a small bunodont anthracothere). The bunodont lower and upper third molars from Caribari are compatible with each other in morphology and dimensions, on which basis it is concluded that they represent a single species, in recent literature known as *Microbunodon silistrensis*.

Because of this discovery, the previous attribution of the Indo-Pakistan small bunodont anthracotheres to the genus *Microbunodon* requires re-assessment. The morphological differences concerning the mandibles of European Oligo-Miocene species and the Miocene ones from Indo-Pakistan evoked by Lihoreau et al. (2004), opens the possibility that there are two or more genera involved. The anterior parts of mandibles from Indo-Pakistan are scarce (Colbert, 1935; Lihoreau et al. 2004) and no skulls have been described, which hinders comparisons between them and the European forms.

Comparison of the augmented sample of Indo-Pakistan small bunodont anthracotheres with specimens from Europe reveals that they differ from *Microbunodon* Depéret, 1908, and *Microselenodon* Depéret, 1908, as suspected by Colbert (1935) who wrote that the Siwalik forms might be included in *Microselenodon* but because the latter genus was poorly defined it was better to leave the Indian fossils in *Anthracotherium*. The newly available samples from Europe (La Milloque) and Indo-Pakistan, especially the undescribed material from Ramnagar which was consulted on Morphosource (Patel et al. 2022), provide additional information that clarifies the affinities of the 'Siwalik' taxa.

The additional material from La Milloque, France (Brunet, 1968; Lihoreau et al. 2004) was for a long time included in *Microbunodon minimum* but, because *Anthracotherium minimum* is the type species of *Microselenodon*, and because it clearly differs in morphology and dimensions from the type species of *Microbunodon* (*Mb. laharpei* (RENEVIER, 1879)) the Milloque species should be classified in *Microselenodon*. The outcome, as suspected by Colbert (1935) is that the 'Siwalik' forms are not particularly close to *Microselenodon*, nor to *Microbunodon*, and they also differ sufficiently from *Anthracotherium* to necessitate the erection of a new genus, herein called *Garobunodon*, named for the Garo Hills from which the type species, previously known as *Anthracotherium silistrense*, was collected.

Metric analysis of the small bunodont anthracotheres from the Miocene deposits (Siwaliks, Bugti, Sind etc.) of Indo-Pakistan reveals the presence of three size groups, the largest one containing the holotypes of two species (*Anthracotherium silistrense* (PENTLAND, 1828) and *Microbunodon milaensis* LIHOREAU, BLONDEL, BARRY & BRUNET, 2004), the medium-sized one containing the holotypes of two species, in the order of naming: *Microbunodon punjabiensis* (LYDEKKER, 1877a) and *Microselenodon mus* (PILGRIM, 1908) and the smallest group contains the lectotype of *Anthracotherium exiguum* FORSTER-COOPER, 1924. It is concluded that the valid names of the species representing the three size groups are: 1) *Garobunodon silistrensis* for the largest of the three species, 2) *Garobunodon punjabiensis* for the medium-sized species, and *Garobunodon exiguus* for the smallest of the three.

Concerning the age of the Caribari sedimentary deposits, the new interpretation of the fossils indicates that the association of two species of small anthracotheres (*Garobunodon silistrensis* and *Sivameryx palaeindicus*) in the Bengali deposits, accords best with an age estimate of Late Middle Miocene to basal Late Miocene (ca 12–9.5 Ma) instead of the traditional view that the deposits were of Early Miocene age.

Material and methods

Fossils, or casts of fossils, were examined wherever possible, except for the material in American institutions described by Lihoreau et *al.* (2004) and the specimens described by Khan et al. (2013) for which data was accepted as published by these authors unless there were doubts about the appurtenance of the specimens. For example, PUPC 13/1, the tooth interpreted by Khan et al. (2013) to be a left m/1 of *Microbunodon silistrensis* is in fact the rear two thirds of a d/4 of a suid, possibly *Conohyus* PILGRIM, 1926, or *Kachchhchoerus* BHANDARI, PICKFORD & TIWARI 2015. Access was also made to data and images in the Morphosource website (Patel et al. 2022) for fossils from Ramnagar, India, and the online portals of the NHMUK and the AMNH for information concerning their discovery locus and collector.

The fossils from Indo-Pakistan were compared with other genera of Microbunodontinae: *Microbunodon* DEPÉRET, 1908, *Microselenodon* DEPÉRET, 1908, *Anthracokeryx* PILGRIM & COTTER, 1916, *Geniokeryx* DUCROCQ, 2020, *Oligokeryx* DUCROCQ, CHAIMANEE, YAMEE & JAEGER, 2025, *Nabotherium* SILEEM, SALLAM, HEWAIDY, MILLER & GUNNELL, 2016 and *Etruscotherium* PICKFORD, 2021.

Specimens were measured with vernier calipers to the nearest tenth of a mm. It is clear that different researchers have measured teeth using different landmarks (see Annex). In this paper, the mesiodistal length (MDL) of upper molars was measured on the buccal side of the teeth and allowance was made for interstitial wear that can affect the extent of the parastyle, and the buccolingual breadth (BLB) was measured across the mesial loph at its broadest part, unlike the method of Lihoreau (2003, fig. II.3) who measured the breadth of the distal loph of upper molars. Inter-observer differences can be significant: for example MDLxBLB of the M3/, YPM 20098, was published as 15.4x16.7 mm by Lihoreau et al. (2004) whereas my own measurements are 16.9x18.1 mm.

Images were captured with a Sony Cybershot digital camera, and the resulting photographs were treated using Photoshop Elements 15 to remove unwanted background, to reduce trembling and to enhance contrast. Scales were added at the treatment stage in order to minimise the effects of parallax.

For the compilation of the synonymy lists the system of Matthews (1973) was employed.

The NHMUK has two catalogues of fossils, one informally known as the Old Register (OR) the other as the 'M' catalogue, and it is clear that the duplication of numbers in the two catalogues has caused a certain amount of confusion among researchers who did not realise the existence of the two catalogues (e. g. Forster-Cooper, 1924 who gave the numbers of the Caribari fossils with 'M' as a prefix, whereas they do not have an alphabetic prefix – they feature in the PV OR catalogue of the NHMUK). For example, a search in the 'M' catalogue for data concerning a fossil registered in the Old Register will yield the wrong information about the fossil, the locality from which it came, the taxon and other archival information about the specimen. In the former catalogue, specimens were numbered without an alphabetical letter before the number, whereas in the latter catalogue, the letter 'M' appears before the number. When publishing on the fossils that feature in the Old Catalogue, the letters OR should be put before the number, even though these letters do not feature on the fossils themselves. The fossils registered in the 'M' catalogue have the letter 'M' on the fossils themselves, followed by the number. Lihoreau et al. (2004) concluded that the fossil that they listed as M19442 was from an unknown locality, but it is numbered without an 'M' before the digits, and it thus features in the Old Register where its location is given as Caribari, once its correct specimen number (19042) has been determined.

Dental nomenclature

Figure 1 explains the nomenclature of the main cusps, minor cusplets and crests of molariform teeth of *Garobunodon*. The cristae and cristids are depicted by solid lines and the grooves between cusps and cusplets by dotted lines. Reference was made to the nomenclature illustrated in Lihoreau & Ducrocq (2007).

Abbreviations

AMNH	American Museum of Natural History, New York
GSI	Geological Survey of India, Calcutta
GSP	Geological Survey of Pakistan, Quetta
IM	Indian Museum, Calcutta (the same place as the GSI)
NHMUK	Natural History Museum of the United Kingdom, London
PUPC	Palaeontology Laboratory, Zoology Department, University of the Punjab, Lahore, Pakistan
PV M	Palaeontology Vertebrata, Mammalia
PV OR	Palaeontology Vertebrata, Old Register
WIHG	Wadia Institute of Himalayan Geology, Dehra Dun, India
YPM	Yale Peabody Museum, Yale University, New Haven, Connecticut, USA

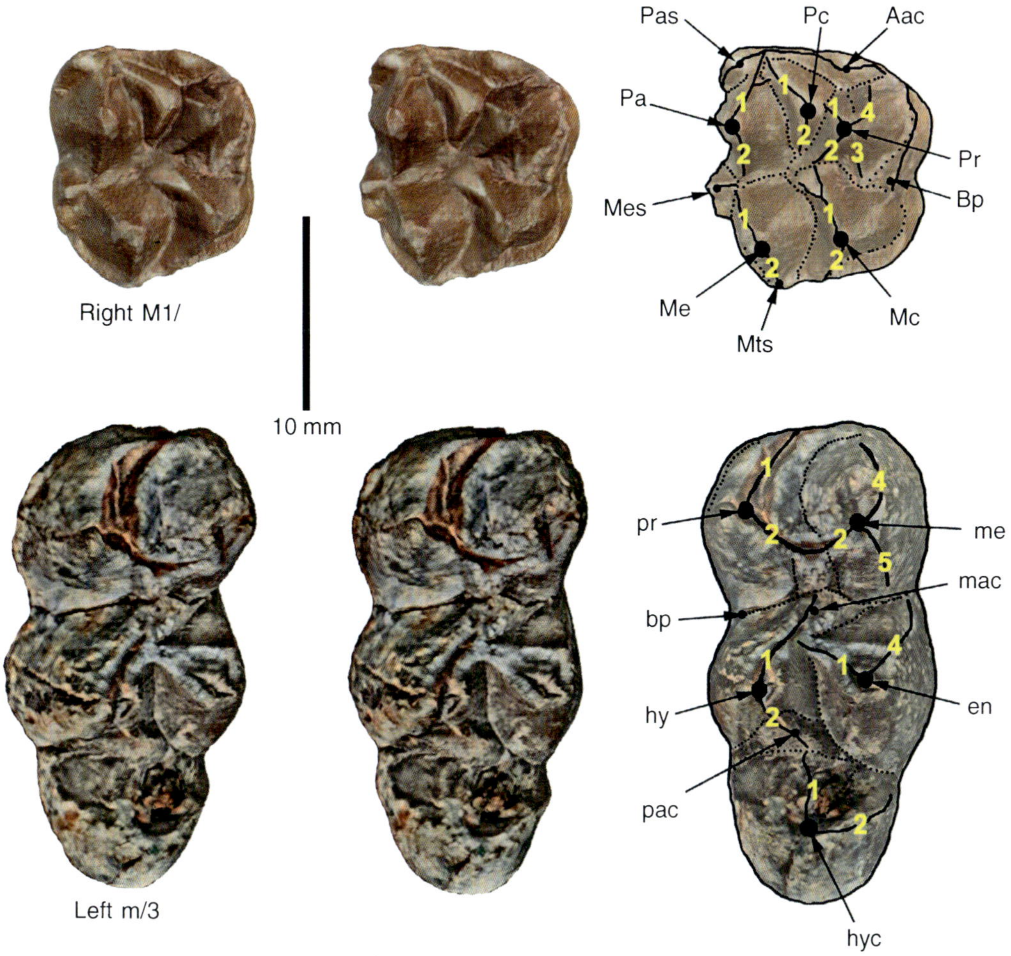

Figure 1. Stereo occlusal views of a right M1/ and a left m/3 of *Garobunodon silistrensis* with the nomenclature of the cusps, cusplets, crista and cristids. Upper molariform: Aac, Anterior accessory cusplet (Protostyle); Bp, Basal pillar; Mc, Metaconule; Me, Metacone; Mes, Mesostyle; Mts, Metastyle; Pa, Paracone; Pc, Paraconule; Pr, Protocone; Pas, Parastyle. Lower molariform: bp, basal pillar; en, entoconid; hy, hypoconid; hyc, hypoconulid, mac, median accessory cusplet, me, metaconid; pac, posterior accessory cusplet; pr, protoconid. (**1,** precrista and precristid; **2,** postcrista and postcristid; **3,** lingual postprotocrista; **4,** mesiolingual protocrista and mesiolingual cristid; **5,** distolingual metacristid). (NB: in Ghezzo & Giusberti (2016, fig. 4) the Protostyle was labelled the Mesostyle, the Mesostyle was labelled the Ectostyle and the basal pillar was labelled the Entostyle).

Taxonomic issues European Microbunodontinae

The most recent taxonomic revision of the small bunodont anthracotheres from the Indian Subcontinent by Lihoreau et al. (2004), contains a number of errors, including the incorrect naming of the type species of *Microbunodon* DEPÉRET, 1908, as well as that of *Microselenodon* DEPÉRET, 1908. The authors wrote that the type species of *Microbunodon* is *Anthracotherium minimum* Cuvier, 1822, yet Depéret (1908) designated *Anthracotherium laharpei* as the type species. He wrote:

"*Deuxième rameau.* – Microbunodon *n. g.; type* Anthracotherium laharpei RENEVIER. *Ce rameau, encore mal suivi dans son évolution, comprend de petites formes à dentition brachybunodonte, comme celles du premier rameau. L'espèce type de Rochette, connue seulement par ses molaires supérieures (M3/=20 mm) portait de longues canines aplaties, traguliformes, qui distinguent bien les* Microbunodon *de tous les autres Anthracothéridés. Il est possible que l'*A. sandbergeri *v.* MEYER, *des lignites aquitaniens de Gusternhain (Westerwald), connu seulement par ses molaires inférieures, appartienne à la même espèce ou à une mutation du même rameau*".

In the same paper, Depéret (1908) designated *Anthracotherium minimum* as the type species of the genus *Microselenodon*. He wrote:

"*Troisième rameau.* – Microselenodon *n. g.; type* Anthracotherium minimum CUVIER. *Rameau formé également de petites espèces, remarquables par le type brachysélénodonte des M, surtout aux denticules internes des M supérieures et aux denticules externes des inférieures. Ces caractères sont probablement corrélatifs de différences importantes du crâne et du squelette. Leur ligne d'évolution est peu connue: le* M. minimum *ne débute que dans le Stampien, et se continue sans changement notable dans l'*A. breviceps TROSCHEL *de l'Aquitanien de Rott (m/3 infér.=20 mm dans le premier, 23 mm dans le second)*".

Stehlin (1910) correctly noted that Depéret (1908) had erected two new genera of small bunodont anthracotheres

(*Microbunodon*, with type species *A. laharpei*, and *Microselenodon* with type species *A. minimum*). However, he decided that there was only one genus represented and transferred *A. minimum* to the genus *Microbunodon*. He incorrectly reported that the genus had triangular upper canines, seemingly confusing the word '*traguliformes*' employed by Depéret (1908) for 'triangular' ('*dreikantigen*').

Pilgrim (1912, pp. 41,46) called the two species of small bunodont anthracotheres from Indo-Pakistan that he studied, *Microselenodon mus* and *Microselenodon silistrensis*. In doing so, he adhered to the designation of *Anthracotherium minimum* as the type species of the genus *Microselenodon* by Depéret (1908). He later attributed both species to *Microbunodon* (Pilgrim, 1913).

Forster-Cooper (1924, p. 18) wrote *"As the genus* Microbunodon *according to Depéret's diagnosis (1) is very sharply marked off from the others by the triangular upper canines and premolariform lower canines and as considerable doubt has been cast (2) on the validity of his genus* Microselenodon, *which is certainly insufficiently characterised, it is necessary to place the following specimens, the front parts of whose skulls are as yet unknown, temporarily in the genus* Anthracotherium *in its widest sense"* (1 = Depéret, 1908; 2 = Stehlin, 1910).

However, both Stehlin (1908) and Forster-Cooper (1924) as well as Pilgrim (1928) misread Depéret's word '*traguliformes*', thinking that he had written 'triangular', and for this reason, the comments about the genus *Microbunodon* being sharply marked off from other anthracotheres are invalid (see also Hellmund, 1991, for descriptions of similar upper canines in the bothriodont anthracothere *Elomeryx crispus* (GERVAIS, 1849)). Furthermore, Forster-Cooper (1924) did not clarify why he considered *Microselenodon* to be insufficiently characterised to be a valid taxon. Depéret (1908) nominated a type species (*Microselenodon minimum*) which means that if the genus *Microselenodon* is insufficiently characterised, then nominating *A. minimum* as the type species of *Microbunodon*, as was incorrectly done by Lihoreau et al. (2004), would mean that *Microbunodon* is also insufficiently characterised. Apart from that, the holotype of *A. minimum* CUVIER, 1824, is a mandible fragment with m/1-m/3 (Cuvier, 1822, p. 404-405, pl. 80, fig. 5) and thus provides no information about the morphology and dimensions of the upper canines.

When he described the canines of the *Microbunodon* and *Microselenodon* as '*triangular*' Pilgrim (1928) propagated the erroneous reading of Depéret's (1908) word '*traguliformes*'. In his differentiation of *Anthracokeryx* from the European genera, Pilgrim (1928) wrote "*Apart from the shortening of the premolar series, the triangular cross-section of the canine and the backward shifting of the inner cusp of p/3, the structure of the upper molars and of the lower molars and premolars is hardly at all different from that of the Pondaung genus*". Correct this misreading, and it becomes evident that the European and Pondaung genera are more similar to each other than was concluded by Pilgrim (1928).

Matthew (1929) summarised the opinion of Pilgrim (1912) but incorrectly wrote that Pilgrim had assigned the species *A. silistrense* and *A. mus*, to *Microbunodon*. As noted above, Pilgrim (1912) attributed both of these species to the genus *Microselenodon*, although a year later (Pilgrim, 1913) he did list both of them as *Microbunodon*.

Colbert (1935) examined the taxonomic issues related to the small bundont anthracotheres from the Indian Subcontinent. He wrote "*Forster-Cooper, in 1924, evidently misinterpreted Depéret's analysis of the genus, because he makes the following statement. 'As the genus* Microbunodon *according to Depéret's diagnosis is very sharply marked off from the others by the triangular upper canines and premolariform lower canines.' Evidently Forster-Cooper mistook the word traguliform for triangular. Then supposing* Microbunodon *to be distinguished by triangular upper canines, he excluded* Anthracotherium silistrense *from it. So we see that* Anthracotherium punjabiense *(called* A. silistrense *by Forster-Cooper) can not be excluded from the genus* Microbunodon *on the grounds that it does not have a triangular canine. There are, however, good reasons for placing* A. punjabiense *in a genus different from* Microbunodon. *In speaking of the species under discussion, Matthew says: 'Pilgrim refers this to* Microbunodon, *but it is remarkably selenodont for an anthracotherine, as much so as* Brachyodus, *etc.' In the original definition of* Microbunodon, *Depéret stipulates that it is a small brachybunodont anthracothere. Consequently it would seem logical to exclude the Siwalik species from Deperets genus, in the light of Matthew's positive observations of its very selenodont teeth*".

Colbert (1935) summed up: "*Of course, the Siwalik form might be included in Depéret's genus* Microselenodon, *as it once was by Pilgrim. But in view of the fact that the definition of this latter genus is not very clear, and that the necessary information for such an inclusion is not just now available, it seems best to place the form under discussion in the genus* Anthracotherium, *recognizing the desirability of a future transference to a more appropriate genus*".

In their differential diagnosis of *Microbunodon*, Lihoreau et al. (2004) wrote: "Anthracotherium *differs from* Microbunodon *by a developed entoconulid, an accessory cusp on upper molar mesial cingulum and on P3/ lingual cingulum, a strong and vertical lower canine in contact with upper canine, a straight mandibular corpus, a flattened ventral border of symphysis, a third hypocristid, on lower molar, that joins pre-entocristid forming a hypolophid, and lower incisors mesiodistally narrowed and high crowned*".

Ducrocq & Lihoreau (2007) wrote that the type species of *Microbunodon* is *Microbunodon minimum* (CUVIER, 1822) but this not so, Depéret (1908) having specified that the type species of his new genus is *Anthracotherium laharpei* RENEVIER. *Anthracotherium minimum* is the type species of the genus *Microselenodon* DEPÉRET, 1908. Apart from that, the nomen *Anthr. minimum* was first printed in 1824.

The fossils from Indo-Pakistan attributed to *Microbunodon* by these authors do not fully comply with their diagnosis. For example, 1) the presence of a '*developed entoconulid*' in the m/3 (not defined in Lihoreau et al. 2004, but defined by Lihoreau, 2003, fig. II-17, character 25) is not apparent in any of the specimens from Indo-Pakistan, nor is it present in *Anthracotherium* (vide Pickford, 2016); 2) all the upper molars of *Garobunodon silistrensis* from Caribari, Ramnagar and Chur Lando possess a well-developed accessory cusplet on the mesial cingulum, and by this feature recall *Anthracotherium* (vide Pickford, 2016) rather than *Microbunodon* (sensu Lihoreau et al. 2004) or *Microselenodon*; 3) '*a third hypocristid, on lower molar, that joins pre-entocristid forming a hypolophid*' (defined by Lihoreau, 2003, fig. II-2 and fig. II-19, character 27) is present in *Anthracotherium* and occurs in all the lower molars of bunodont anthracotheres from the Indian Subcontinent.

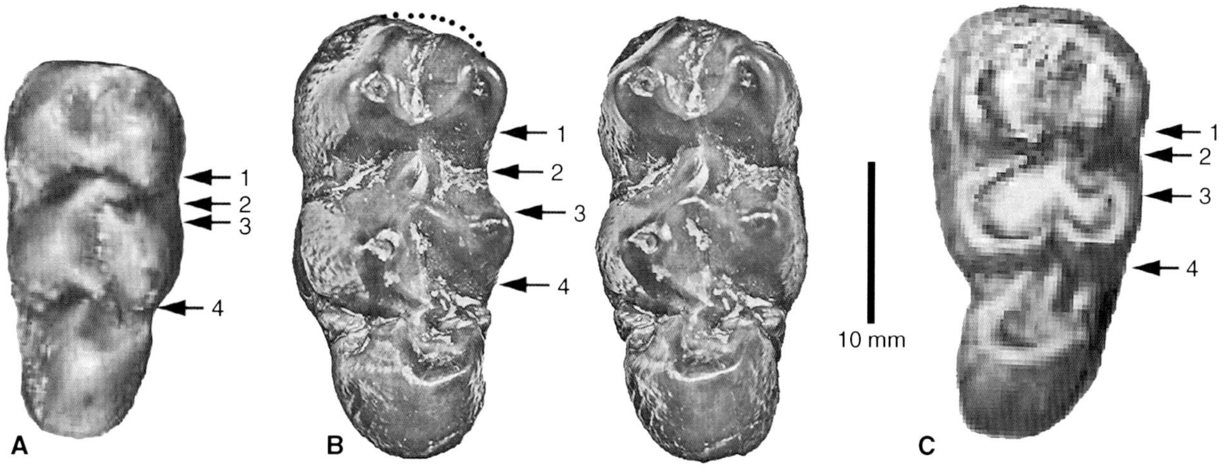

Figure 2. Comparison of left occlusal views of m/3s of European and Indo-Pakistani small bunodont anthracotheres. **A,** *Microselenodon minimum* from La Milloque (image modified from Lihoreau et al. 2004), **B–C,** *Garobunodon silistrensis*; **B,** stereo pair (reversed) of NHMUK PV OR 19042 from Caribari; **C,** holotype of *Microbunodon milaensis*, YPM 10999 from Dhok Mila, image modified from Lihoreau et al. 2004). Arrows show the main differences between teeth of *Microselenodon* and *Garobunodon*: **1,** distolingual metacristid well-developed in *Microselenodon*, weak in *Garobunodon*; **2,** mesial end of prehypocristid transversely oriented in *Microselenodon*, not twisted in *Garobunodon*; **3,** preentocristid pointing towards the apex of the protoconid in *Microselenodon*, pointing towards the distal base of the protoconid in *Garobunodon*; **4,** distolingual entocristid well developed in *Microselenodon*, absent in *Garobunodon*.

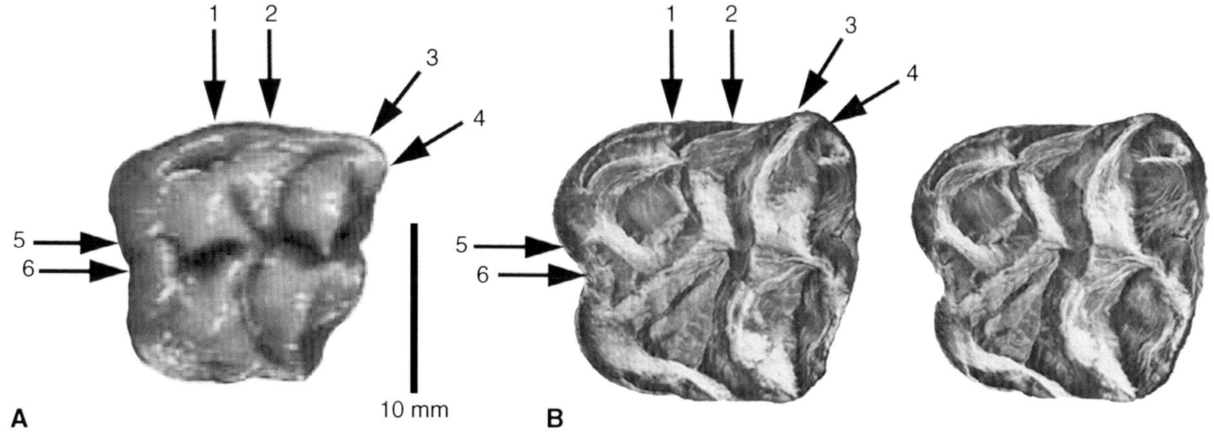

Figure 3. Comparison of left upper third molars of *Microselenodon minimum* (**A**, from La Milloque, image modified from Lihoreau *et al.* 2004) and *Garobunodon silistrensis* (**B**, stereo pair (reversed) from Ramnagar). Arrows show the main differences between the teeth of the two genera: **1,** anterior accessory cusplet weak in *Microselenodon*, well-developed in *Garobunodon*; **2,** postparaconule cristule not reaching median transverse valley in *Microselenodon*, broad and extending distally to contact the premetaconule cristule in *Garobunodon*; **3,** groove between paracone and paraconule markedly curved buccally towards its mesial end in *Microselenodon*, not curved buccally in *Garobunodon*; **4,** parastyle buccally positioned in *Microselenodon*, mesially positioned in *Garobunodon*; **5,** postprotocrista elongated and extending into the median transverse valley in *Microselenodon*, short and terminating far from the median transverse valley in *Garobunodon*, + lingual postprotocrista lingually positioned in *Microselenodon*, more internally positioned in *Garobunodon*; **6,** cusplet in the lingual end of the median transverse valley weak in *Microselenodon*, strong in *Garobunodon*.

Because no canines, nor incisors, of small bunodont anthracotheres have been described from Indo-Pakistan, it is not possible to make comparisons with the specimens from Europe attributed to *Microbunodon minimum* by Lihoreau et al. (2004) that do contain these teeth.

The attribution of the small bunodont anthracothere specimens from Indo-Pakistan to *Microbunodon* by Lihoreau et al. (2004) is not supported by the morphological evidence. The attribution of some of the fossils to *Microselenodon* by Pilgrim (1908, 1912) obtains little, if any, support because the molars in the holotype mandible fragment of the type species *Microselenodon minimum* (Cuvier, 1824) differ from the specimens from the Indian Subcontinent in several features (Figs 2, 4) and there are clear differences between the upper third molars of *Microselenodon minimum* and *Garobunodon silistrensis* (Figs 3, 5).

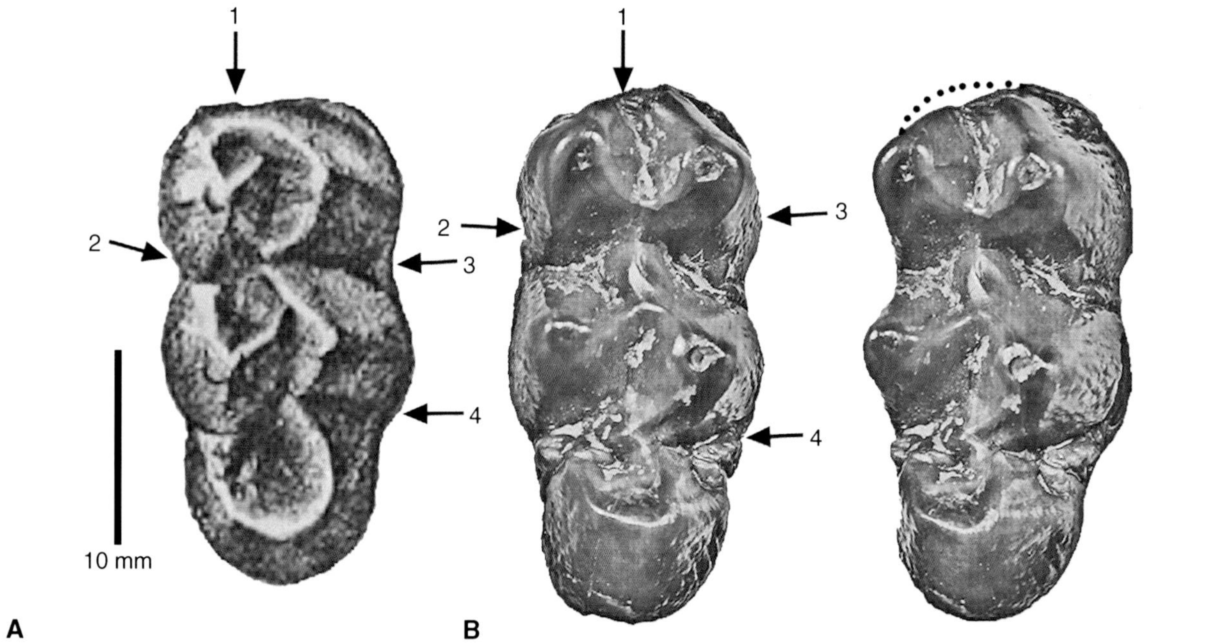

Figure 4. Comparison of right occlusal views of m/3s of *Microbunodon laharpei* from Rochette (**A,** image reversed from Kowalevsky, 1874) and *Garobunodon silistrensis* from Caribari (**B,** stereo pair). The main differences between these teeth are: **1,** premetacristid present in *M. laharpei*, absent in *G. silistrensis*; **2,** postmetacristid moderately well-developed in *Mb. laharpei*, weak in *G. silistrensis*; **3,** postprotocristid elongated in *Mb. laharpei*, shorter and more transversely oriented in *G. silistrensis*; **4,** hypoconid more mesially positioned than entoconid in *Mb. laharpei*, more distally positioned in *G. silistrensis*.

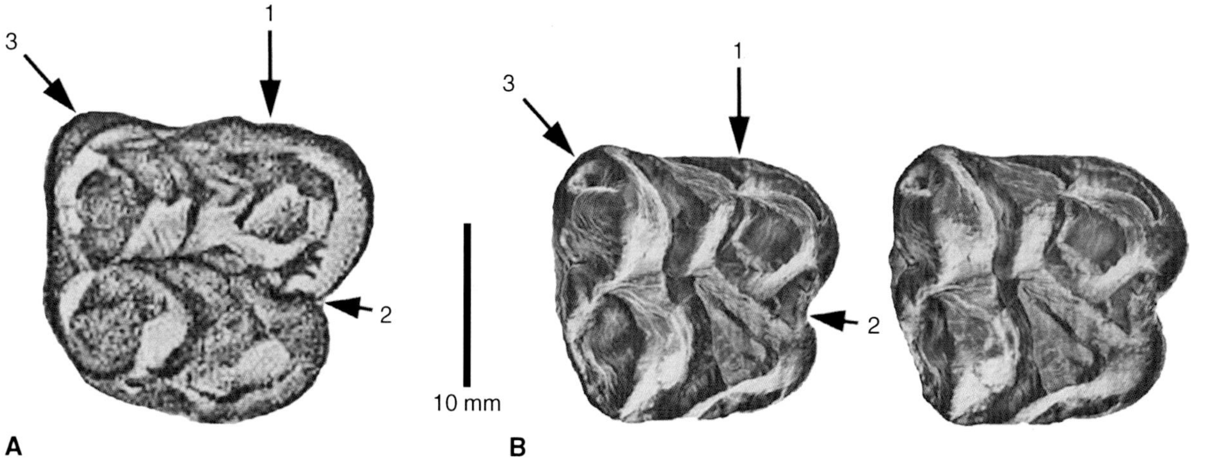

Figure 5. Comparison of occlusal views of right upper third molars of *Microbunodon laharpei* (**A,** image of the holotype (reversed) modified from Kowalevsky, 1874) and *Garobunodon silistrensis* (**B,** from Ramnagar, stereo pair). The arrows show the main morphological differences between the teeth: **1,** anterior accessory cusplet (protostyle) not individualised in *Mb. laharpei*, well-developed in *G. silistrensis*; **2,** accessory cusplet at lingual end of median transverse valley weakly differentiated in *Mb. laharpei*, strongly developed in *G. silistrensis*; **3,** parastyle strong and projecting in *Mb. laharpei*, weaker in G. *silistrensis*.

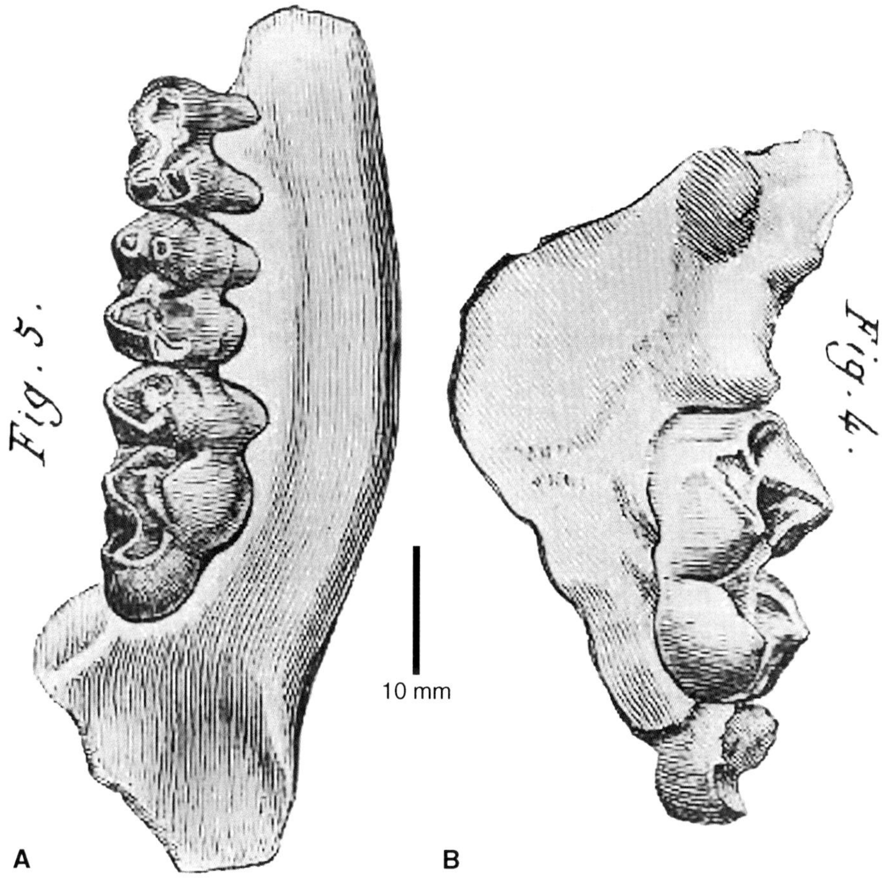

Figure 6. Illustrations of the type specimens of *Anthr. minimum* (**A**) and *Anthr. minus* (**B**); **A,** left mandible fragment with m/1-m/3 from Hautevigne (Lot-et-Garonne); **B,** right mandible fragment with m/3 from Cadibona (Italy) (drawings modified from Cuvier, 1822).

Small bunodont anthracotheres from the Oligo-Miocene of Europe

Following the publication by Stehlin (1910) in which he proposed that the two genera erected by Depéret (1908) were synonyms, most authors have accepted the line priority of *Microbunodon* over *Microselenodon*. However, the taxonomic lumping of all the species hitherto attributed to these genera has not resulted in consensus. Some authors (Goldi, 1914; Hünermann, 1967) have taken the line priority of *Anthr. minus* Cuvier, 1824 (often cited as 1822) over *Anthr. minimum*, as the basis for determining the valid species name, despite the poor state of the type specimen and the slightly oblique image of it that renders it difficult to interpret. The type specimen of *Anthr. minus* (Fig. 6B) most probably belongs to *Anthracotherium* (see below) and thus its transference to *Microbunodon* is not sustainable.

On the grounds that the type specimen of *Anthr. minimum* (Fig. 6A) is better preserved and more informative than the type specimen of *Anthr. minus*, and that other specimens from the same region in France clarify the taxonomic status of the former species, some authors (Brunet, 1968; Lihoreau et al. 2004) have proposed that it should be called *Microbunodon minimum*. However, this species was designated as the type species of *Microselenodon* by Depéret (1908). Furthermore, Filhol (1891a, p. 91) stressed that, even though *Antracotherium* (sic) *minimum* from La Milloque shared some resemblances with material from the lignites at Rott and Rochette (i. e. *Anthr. laharpei*), the latter specimens are clearly distinguished '*by a series of peculiarities*' from those collected at La Milloque (he did not provide details). Thus, maintaining the placement of *A. minimum* in the genus *Microbunodon* is not a stable solution to the systematics or taxonomy of these small European anthracotheres.

The type specimen of *Microbundon minus* (Cuvier, 1824, plate 80, fig. 4), from Cadibona, Italy, was said to be an m/3 by Cuvier (1822, p. 413). He wrote "*C'est une dernière molaire entièrement semblable à son analogue dans la grande espèce, si ce n'est que son dernier tubercule est plus profondément bifurqué, et que ses deux lobes ne sont pas entièrement à cote l'un de l'autre. Sa longueur est de 0,033; sa largeur en avant de 0,013. Ainsi elle a moitié de l'autre espèce en longueur mais beaucoup moins en largeur; elle est donc plus étroite à proportion*".

Hünermann (1967) accepted Cuvier's (1822) interpretation of the Cadibona tooth as an m/3. He considered that, because *Anthr. minus* has line priority over other names of small bunodont anthracotheres from Europe, it is therefore the senior name for them. However he transferred the species to *Microbunodon* as the combination *Mb. minus*. He listed *Mb. minimum*, *Mb. sandbergeri*, *Mb. breviceps* and *Mb. laharpei* as synonyms of *Mb. minus*. However, such an attribution is not without its difficulties because 1) the tooth is too large to be attributed to any of the small bunodont anthracotheres known from Europe and Indo-

Pakistan, and 2) its hypoconulid is bifurcate, a character that does not occur in *Microbunodon*, *Microselenodon* or *Garobunodon* but which is a common feature in species of *Anthracotherium* (Kowalevsky, 1874; Ghezzo & Giusberti, 2016).

The somewhat oblique orientation of the holotype in the illustration of *Anthr. minus* (CUVIER, 1822, pl. 80, fig. 4) renders it difficult to interpret, but what can be made out accords in morphology with an m/3 of *Anthracotherium monsvialense* from Monteviale, illustrated by Ghezzo & Giusberti (2016, fig. 10b). With a length of 33 mm it is small for *Anthracotherium* from Europe, but it is much larger than any specimens attributed to *Microbunodon* or *Microselenodon*. It is considerably longer than the corresponding tooth in *Anthr. minimum* in which the m/3 in the holotype mandible is only 20 mm long. It is also longer than the longest m/3 of *Garobunodon silistrensis* (27 mm long). Its dimensions accord with m/3s of *Anthracotherium crassum*, PILGRIM & COTTER, 1916, from Pondaung, Myanmar (m/3 length ranges from 32.4-39.5 mm (n=9) according to Ghezzo & Giusberti, 2016). It is herein concluded that the species *Anthr. minus* is better left in *Anthracotherium*.

Goldi (1914) wrote that the '*Molasse à lignite*' at Rochette near Lausanne yielded *Anthracotherium* (*Microbunodon*) *minus* despite the fact that this is the type locality of *Microbunodon laharpei*, which is the type species of the genus *Microbunodon*.

Hünermann (1964, 1967) accepted Stehlin's (1910) proposal that there was but a single genus of small bunodont anthracothere in Europe, and he attributed fossils from Ebnat-Kappel to *Microbunodon minus* (Cuvier, 1822) selecting this name on the grounds that it has priority over all other names linked with the genus *Microbunodon*. Apart from the m/3 from Cadibona, the specimens attributed to *Microbunodon minus* by Hünermann (1967) accord better with *Microbunodon laharpei* and *Microselenodon minimum*.

The taxonomic 'lumping' by Lihoreau et al. (2004) of all the European specimens into a single species, *Microbunodon minimum*, is open to re-evaluation, as there are probably at least two distinct species (and genera) involved: *Ms. minimum* (CUVIER, 1824) and *Mb. laharpei* (RENEVIER, 1879) if not more (*Mb. sandbergeri* (VON MEYER, 1852) and *Mb. breviceps* (TROSCHEL, 1859) (Boettger, 1877).

Metric analysis European Microbunodontinae

The available metric data concerning small bunodont anthracotheres from Europe, when taken together with the geographic distribution of the fossils, suggests strongly that there are at least two species in the European samples. This is because several sites have yielded remains only of the larger of the two metric groups (*Microbunodon laharpei*), whereas the smaller of the two metric groups (*Microselenodon minimum*) is rare except at La Milloque, where the small specimens (38 m/3s and ten M3/s) greatly outnumber the large ones (six m/3s and one M3/).

The question as to whether or not there are two genera of small bundont anthracotheres in the Oligo-Miocene of

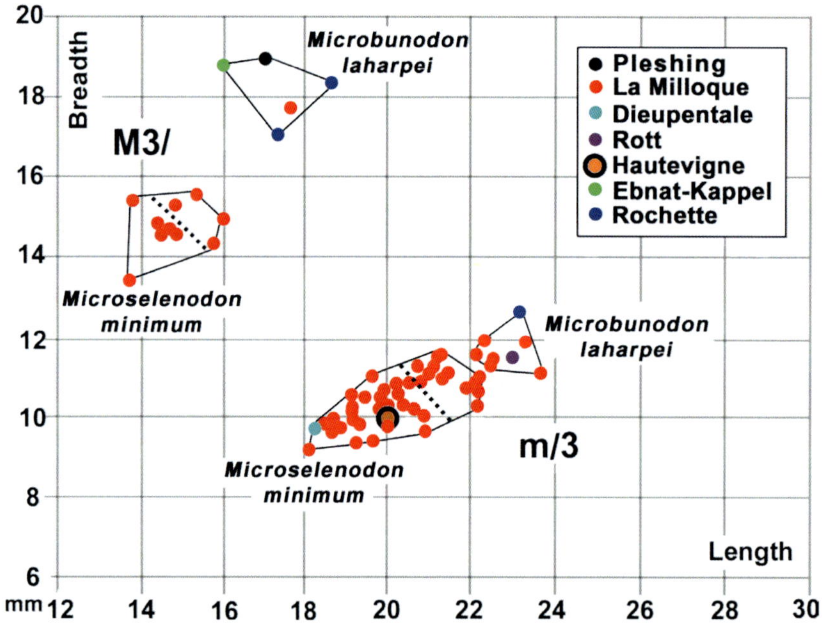

Figure 7. Bivariate (length-breadth) plots of upper and lower third molars of *Microbunodon* from Europe (employing the data points in Lihoreau et al. 2004, fig. 4). Added to the graph are an m/3 from Rochette, and an M3/ from Pleshing (near Linz, Austria) the latter tooth attributed to *Microbunodon* cfr *minus* by Sickenberg (1934, p. 60, specimen N° 27/1929). The M3/s plot into two well-separated clusters; the m/3s into two or possibly three clusters (two are shown here, but the cluster of smaller specimens could be subdivided into two: the dotted lines). Highlighted is the m/3 from the type locality of *Microselenodon minimum* (Hautevigne). Rott is the type locality of *Mb. breviceps*, and Rochette is the type locality of *Mb. laharpei*.

Europe remains open, but the currently available metric data (Fig. 7) indicate that there are at least two species present. The morphological differences between the teeth of *Microbunodon laharpei* and *Microselenodon minimum*, lend support the interpretation that the two species belong to two distinct genera.

In their metric analysis, Lihoreau et al. (2004, fig. 4) included fossils that they attributed to *Microbunodon minimum* from six localities in Europe. The vast majority of specimens in their analyses came from La Milloque with fewer specimens from Dieupentale, Rott, Hautesvignes, Ebnat-Kappel and Rochette (material from a diversity of other localities in Europe did not feature in the metric analysis). The breadth of the upper molars was of the second loph, which is narrower than the mesial loph. Their measurements differ somewhat from those published by other authors (eg Hünermann, 1967).

The included fossils had previously been attributed to different species (for details, see synonymy list in Lihoreau et al. 2004) including *Anthracotherium sandbergeri*, *Anthracotherium breviceps*, *Anthracotherium laharpei* and *Microbunodon minus* (see Hünermann, 1967). The bivariate plots of the upper and lower third molars (Fig. 7) of these European fossils indicate that more than one species is probably present in the hypodigm of *Mb. minimum* sensu Lihoreau et al. (2004). For example, the upper third molars from Rochette, Pleshing and Ebnat-Kappel, plot outside the range of variation of the main cluster of points of *Mb. minimum* from La Milloque, with only one specimen from the latter site being almost as large as them. The m/3 from the type locality of *Ms. minimum* (Hautevigne) plots close to the centre of the cluster formed by the Milloque specimens, which are thus reasonably attributed to the species, but the specimens from Rochette, Pleshing and Ebnat-Kappel, probably do not belong to this small species, but to a larger one that has dimensions similar to *Garobunodon silistrensis*.

Stehlin (1910) concluded that there were two (or possibly three) species of small bundont anthracotheres in the European samples, a finding supported by the present analysis.

A revised taxonomy of the European forms accepting the original designation of type species by Depéret (1908), and taking into account the metric data, is provided below.

Genus *Microbunodon* Depéret, 1908

Type species: by original designation; *Anthracotherium laharpei* Renevier, 1879

Type locality: Rochette, Switzerland

Other species: *Mb. sandbergeri* (von Meyer, 1852), *Mb. breviceps* (Troschel, 1859). The literature indicates the potential for these taxa to be synonyms of the type species.

Discussion

Lihoreau et al. 2004 wrote that the type species of *Microbunodon* is *Anthracotherium minimum* Cuvier, 1822, but Depéret (1908) designated *Anthracotherium laharpei* Renevier as the type species, as was reiterated by Stehlin (1910). *A. minimum* is the type species of the genus *Microselenodon* Depéret, 1908. The type specimen of *Anthracotherium minimum* was illustrated and briefly described by Cuvier 1822 (p. 404, pl. 80, fig. 5) but was not formally named in that publication, but two years later was included in the list of contents of the monograph as *Anthr. minimum* (Cuvier 1824, part ii, p. 528). Thus the date of publication of the nomen *Anthracotherium minimum* is 1824, and not 1822.

Genus *Microselenodon* Depéret, 1908

Type species: by original designation; *Anthracotherium minimum* Cuvier, 1824.

Species *Microselenodon minimum* (Cuvier, 1824)

Type locality: Hautevigne, France (Hautesvignes in Brunet, 1968).

Discussion

Lihoreau et al. (2004) wrote that this species is the type species of *Microbunodon*, in contravention of its designation by Depéret, 1908, as the type species of the genus *Microselenodon*.

Overview of the Microbunodontinae Lihoreau & Ducrocq 2007

The subfamily Microbunodontinae was erected by Lihoreau & Ducrocq (2007) who included two genera within it: *Microbunodon* DEPÉRET, 1908, and *Anthracokeryx* PILGRIM & COTTER, 1916. Several new genera of anthracotheres have been described since then that comply more closely with these two taxa than with genera classified in Anthracotheriinae and Bothriodontinae, so, they are herein classified in the Microbunodontinae. They comprise *Nabotherium* SILEEM, SALLAM, HEWAIDY, MILLER & GUNNELL, 2016; *Geniokeryx* DUCROCQ, 2020; *Etruscotherium* PICKFORD, 2021; and *Oligokeryx* DUCROCQ, CHAIMANEE, YAMEE & JAEGER, 2025. One of Depéret's (1908) genera (*Microselenodon* DEPÉRET, 1908) is added to the subfamily. Also included is the new genus *Garobunodon* diagnosed herein.

Table 1 provides a list of the taxa included in the Microbunodontinae, together with an abbreviated list of synonyms. Further research is required to compile a complete list of all the synonyms. For example, Tsubamoto et al. 2002 considered *Anthracokeryx* to be synonymous with *Anthracotherium,* thereby differing from Lihoreau et al. 2004, who considered it to differ from *Anthracotherium* at the subfamily level.

The species *Anthracokeryx moriturus* Pilgrim, 1928, is of uncertain subfamilial status, but it is probably closer to Anthracotheriinae than to Microbunodontinae (e. g. hypoconulid of m/3 is subbdivided into two well-separated cusplets; Colbert 1938, fig. 52) and was considered to be a synonym of *Anthracotherium crassum* by Tsubamoto et al. 2002. The same uncertainty about subfamilial affinities applies to *Anthracokeryx* (?) *lahirii* Pilgrim, 1928, which was transferred to the Helohyidae, and renamed *Pakkokuhyus* Holroyd & Ciochon, 1995. However, Ducrocq (2019) considered that the latter genus was not a helohyid but a dichobunid. *Anthracokeryx litangensis* Zong, Chen, Huang & Xu, 1996, from China, was transferred to Bothriodontinae by Tsubamoto et al. 2002.

Lihoreau et al. (2004) included in *Microbunodon* the fossil tooth from Masatly, Turkey, illustrated by Newton (1904, pl. 24, fig. 20). Masatly possibly corresponds to modern-day Maksutlu in Edirne Province (Sarigül, 2021). However, the specimen is a bothriodont, as shown by the fact that the posthypocristid closes off the distal fovea on its lingual side, and the prehypocristid terminates closer to the lingual edge of the tooth than in *Microbunodon*, among other features. Morphologically, this tooth is closer to *Merycops* Pilgrim, 1910a, than it is to any microbunodontine.

Table 1. Contents of the anthracotherian subfamily Microbunodontinae arranged by chronological order of naming of the genera and species (type species in bold; synonyms in brown). Homonyms are not listed.

Microbunodon DEPÉRET, 1908
- ***Mb. laharpei*** (RENEVIER 1879)
  - *Mb. sandbergeri* (VON MEYER, 1852)
  - *Mb. breviceps* (TROSCHEL, 1859)

Microselenodon DEPÉRET, 1908
- ***Ms. minimum* (CUVIER, 1824)**

Anthracokeryx PILGRIM ET COTTER, 1916
- ***Ak. birmanicus*** PILGRIM ET COTTER, 1916
- *Ak. tenuis* (PILGRIM ET COTTER, 1916)
  - *Ak. ulnifer* PILGRIM, 1928
  - *Ak. bambusae* PILGRIM, 1928
  - *Ak. myaingensis* PILGRIM, 1928
- *Ak. hospes* PILGRIM, 1928
- *Ak. sinensis* (ZDANSKY, 1930)
  - *Ak. ambiguous* (ZDANSKY, 1930)
  - *Ak. minima* (XU, 1962)
  - *Ak. kwangsiensis* QIU, 1977
  - *Ak. dawsoni* WANG, 1985 (*dawsonae*)
  - *Ak. naduongensis* DUCROCQ, et al 2015

Nabotherium SILEEM, SALLAM, HEWAIDY, MILLER & GUNNELL, 2016
- ***Na. aegyptiacum*** (ANDREWS, 1906)

Geniokeryx DUCROCQ, 2020
- ***Ge. thailandicus*** (DUCROCQ, 1999)
- *Ge. gungkangensis* (QIU, 1977)

Etruscotherium PICKFORD, 2021
- ***Et. ribollaense* PICKFORD, 2021**

Oligokeryx DUCROCQ, CHAIMANEE, YAMEE & JAEGER, 2025
- ***Ok. khiansaensis*** DUCROCQ, CHAIMANEE, YAMEE & JAEGER, 2025

Garobunodon gen. nov.
- ***Ga. silistrensis*** (PENTLAND, 1828)
  - *Mb. milaensis* LIHOREAU et al. 2004
  - *Ms. mus* (PILGRIM, 1908)
- *Ga. punjabiensis* (LYDEKKER, 1877a)
- *Ga. exiguus* (FORSTER-COOPER, 1924)

Provenience

Particular attention was paid to the provenience of the Indo-Pakistan fossil specimens, because the locality data of several of them have been misreported: Lihoreau et al. 2004 wrote that Y 20098 (M3/) was either from an unknown locality (their list of 'new material' on p. 104) or that it was from Chinji (their table 3 on p. 106) but it was collected at Ramnagar, Jammu & Kashmir, by de Terra & Teilhard de Chardin. Several fossils from the Potwar Plateau, Pakistan, were reported, by the same authors, to be from the 'Siwalik Hills', but these hills are in India, over 1000 km southeast of the Potwar Plateau, and the sediments exposed there are mostly Plio-Pleistocene. Two other specimens were listed by Lihoreau et al. (2004) as being from unknown localities, but consultation of the catalogues at the NHMUK reveals that one of them (OR 19042, not M 19442 as listed by the authors) was collected at Caribari, India, and that the other one (M 12707) was collected from Chur Lando, Pakistan. According to the register, NHMUK PV M 19442 is a left mandibular ramus of *Rhinoceros etruscus* from the [Middle Pleistocene] upper Freshwater Bed (loamy sand on top of the black bed) at West Runton, Norfolk, and is part of the A.C. Savin collection purchased in 1946. NHMUK PV OR 19442 is described in the register as a molar in a lower jaw of *Aceratherium incisivum* from [the Miocene of] Darmstadt, Germany, and was purchased from Dr Kaup in July, 1845.

Specimen AMNH 94625 was variably reported by the same authors to be from the 'Siwalik Hills' (their table 3) and from Ramnagar (their list of '*other material*' on p. 104, attributed to *Microbunodon silistrensis*) but according to Lihoreau (2003) it was collected at Ramnagar. The AMNH catalogue lists the specimen as '*48, Ramnagar, Kashmir*' (Novacek & Mehling, pers. comm. 2025). In the same paper, specimen GSI B 104, the holotype of *Anthracotherium punjabiense* was listed as being from the 'Siwalik Hills', but it was collected from the Salt Range (possibly near Hasnot) in the Potwar Plateau (Lydekker, 1878; Pilgrim, 1912).

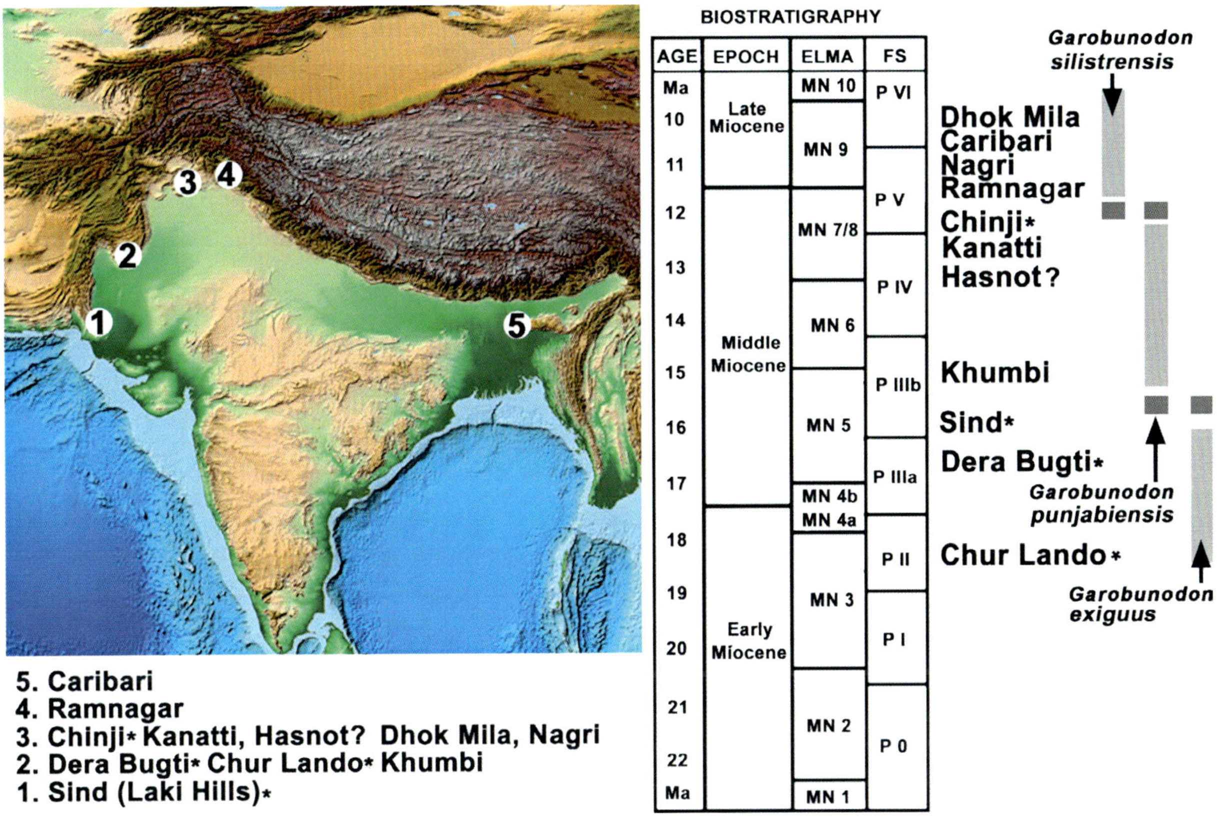

Figure 8. Localities in Indo-Pakistan that have yielded specimens of *Garobunodon* with estimated biostratigraphic ranges of the three species. Hasnot is included with a question mark, even though Lydekker (1878) mentioned this toponym when he described *Anthracotherium punjabiense*. Lydekker (1877a) and Pilgrim (1912) indicated that the location was imprecise, and estimated that the fossil so labelled was probably from the Lower Siwaliks of the Salt Range. Localities with star symbols (Chinji, Sind, Chur Lando and Dera Bugti) expose great thicknesses of strata that span appreciable periods of geological time and have thus yielded remains of more than one species of *Microbunodon* (ELMA – European Land Mammal Ages: Steininger, 1999; FS: Faunal Sets of East Africa: Pickford, 1981, 1988; Pickford & Senut, 2003).

Time scale

The geological time scales employed by diverse authors over the past two centuries have changed dramatically, as was explained by Pickford (1988) and Pickford & Chaïd-Saoudi (2024). The term 'Pliocene', for example, as used by Lydekker (1877a) and Pilgrim (1912) spanned a different period of geological time from that covered by the present day definition of the term. In effect, for many years the term Pliocene, when used for continental faunas (often called Pontian), included the period that is today equivalent to the basal half of the Late Miocene. Thus, when Lydekker (1877a) wrote that *Anthracotherium punjabiense* was from the Pliocene of the Salt Range, he did not mean that it was from deposits younger than the Miocene in today's definition of the latter term (i.e. less than 5.5 Ma) but that it was from deposits that were at the time usually correlated to the Pontian, which, as used for continental faunas, extends back as far as 11.5 Ma (Pickford & Chaïd-Saoudi, 2024). For the time scale of the Indo-Pakistan succession, reference is made to the works by Pickford (1988), Barry & Flynn (1990), Barry et al. (2002) and (Patnaik, 2016).

Localities and age

Dentognathic remains of small bunodont anthracotheres have been collected from about a dozen localities in Indo-Pakistan, ranging in age from the Early Miocene to the basal Late Miocene (Fig. 8). There remains a certain amount of doubt about the precise ages of some of the fossil material collected before the 1970's, in that the geological successions exposed at the localities span considerable periods of geological time (Pickford, 1988; Barry et al. 2002). Nevertheless, a sense of the succession of palaeontological events emerges from the data, even if the precise timing of events is somewhat hazy (Forster-Cooper, 1924; Colbert, 1935; Patnaik, 2016).

Systematic descriptions

Order Artiodactyla OWEN, 1848

Family Anthracotheriidae LEIDY, 1869

Subfamily Microbunodontinae LIHOREAU & DUCROCQ, 2007

Genus *Garobunodon* nov.

Type species: *Anthracotherium silistrense* PENTLAND, 1828

Etymology: Garo for the 'Garo' Hills, Bengal, + 'buno' from βουνός Greek (bounos) for 'hill, mound' in reference to the Garo Hills as well as to the mound-like shape of the molar cusps + 'don', from ὀδούς Greek (odous) for 'tooth'. Garo is an ethnic group that lives in the Garo Hills and neighbouring areas of India and Bangladesh; their language is Garo.

Diagnosis: Small bunodont anthracotheres with a clear anterior accessory cusplet (protostyle) in upper molars on the mesial cingulum opposite the groove between the paraconule and protocone; prominent accessory cusplet at lingual end of median transverse valley in upper molars; postprotocrista in upper molars short and of low relief; postparaconule cristule enlarged and elongated, contacting the premetacristule in the M2/ and M3/; the parastyle in the upper molars is less well developed than those of *Microbunodon, Microselenodon*, *Anthracokeryx, Nabotherium*, *Geniokeryx* and *Oligokeryx*; metaconule in upper molars without ectometacristule; lower molars with low-relief to absent premetacristid ('7' in Lihoreau & Ducrocq, 2007), weak mesial ('5') and distal ('6') metacristids and low-relief rounded mesiolingual ('9') and distolingual ('10') entocristids. Mandibular symphysis is fused, more elongated and not as downwardly projecting as it is in the unfused, short symphysis of *Geniokeryx.*

Species *Garobunodon silistrensis* (PENTLAND, 1828)

Synonymy

pv*1828 *Anthracotherium silistrense* sp. nov. PENTLAND, pp. 393-394, pl. 24, fig. 4-5
pv1847 *Anthracotherium silistrense* PENTLAND; Falconer & Cautley, pl. 68, fig. 23,23a
pv1848 *Choeromeryx silistrense* (PENTLAND); Pomel, p. 687
pv1848 *Anthracotherium silistrense* PENTLAND; Blainville, p. 142
pv1857 *Anthracotherium silistrense* PENTLAND; Rutimeyer, p. 5,9
v1868 *Anthracotherium silistrense* PENTLAND; Falconer, p. 508, fig. 23
pv1877a *Choeromeryx silistrense* (PENTLAND); Lydekker, p. 77
1878 *Anthracotherium silistrense* LYDEKKER; Lydekker, p. 77
1883 *Anthracotherium silistrense* PENTLAND; Lydekker, pp. 149-150, pl. 23, fig. 10,12, (non pl. 24, fig. 1,1a)
1884 *Anthracotherium silistrense* PENTLAND; Teller, p. 53
1885a *Anthracotherium silistrense* PENTLAND; Lydekker, p. 73
1885b *Anthracotherium silistrense* PENTLAND; Lydekker, p. 42
p1904 *Choeromeryx silistrense* (PENTLAND); Trouessart, p. 652
1908 *Anthracotherium silistrense* PENTLAND; Pilgrim, p. 161
non1910a *A.* (*Microselenodon*) *silistrense* (PENTLAND); Pilgrim, p. 201
non1912 *Microselenodon silistrensis* (PENTLAND); Pilgrim, pl. 14, fig. 2
non1913 *Microbunodon silistrense* (PENTLAND); Pilgrim, pp. 285,317
v1924 '*Anthracotherium*' *silistrense* PENTLAND; Forster-Cooper, pp. 18-20, figs 15,16
1924 *Anthracotherium punjabiense* LYDEKKER; Forster-Cooper, p. 20, fig. 17
1929 *Microbunodon silistrense* (PENTLAND); Matthew, p. 453
1929 '*Anthracotherium*' *silistrense* PENTLAND; Matthew, p. 463
?1978 *Anthracotherium punjabiense* LYDEKKER; Vasishat et al., p. 133
1984 *Anthracotherium silistrense* PENTLAND; Raza & Meyer, p. 51
pv1987 *Anthracotherium silistrense* PENTLAND; Pickford & Rogers, p. 645
1987 *Anthracotherium silistrense* PENTLAND; Pickford, pp. 313-316
pv2004 *Microbunodon silistrensis* (PENTLAND); Lihoreau et al. 2004, pp. 103-107, fig. 5, tab. 3
pv2004 *Microbunodon milaensis* sp. nov. LIHOREAU, BLONDEL, BARRY & BRUNET, pp. 107-109, fig. 6, tab. 3

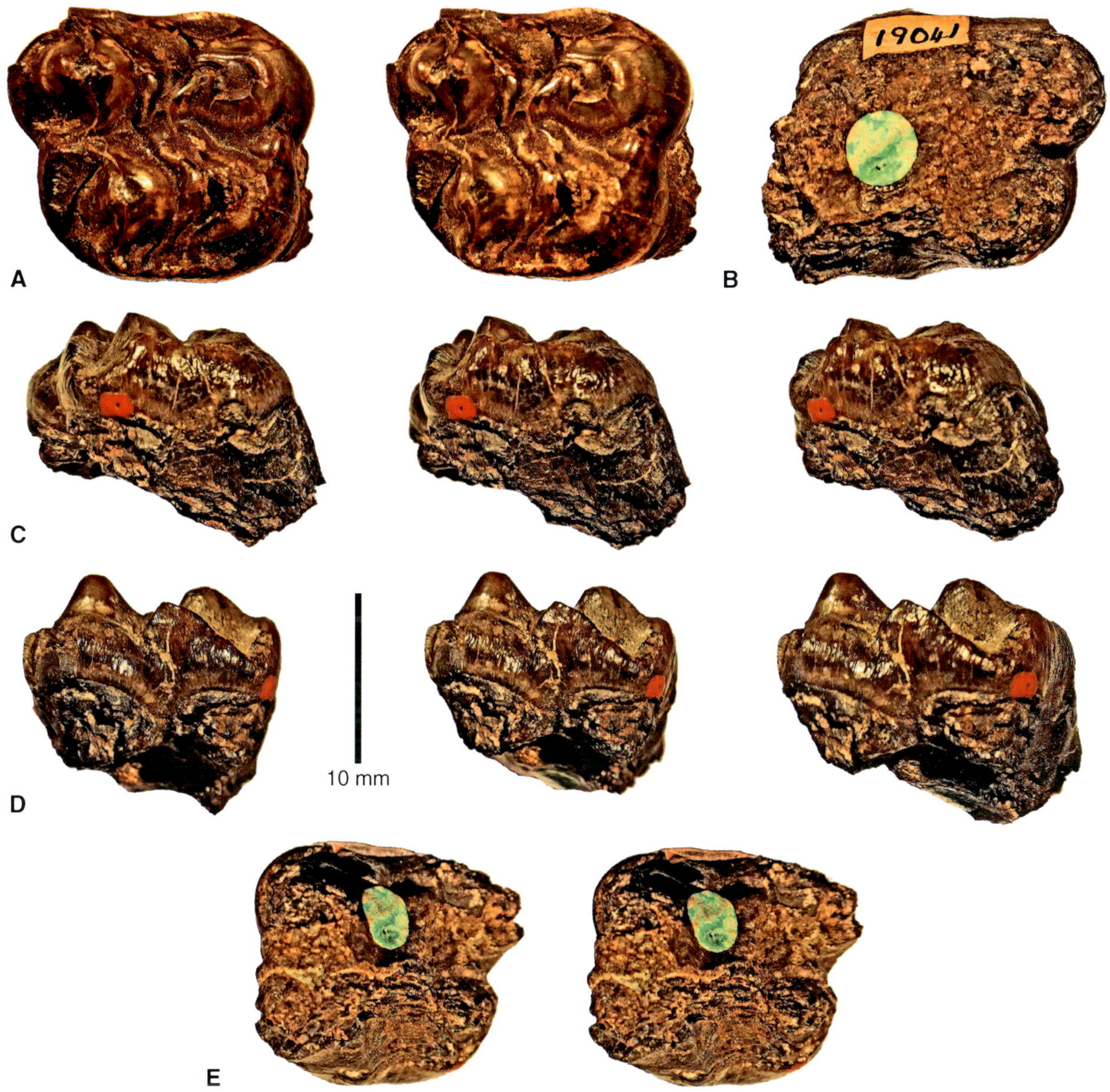

Figure 9. NHMUK PV OR 19041, right M3/, lectotype of *Garobunodon silistrensis* from Caribari, Garo Hills, Bengal, India. **A,** stereo occlusal view; **B,** oblique radicular view to show the specimen number without the prefix 'M'; **C,** stereo distal views; **D,** stereo buccal views to show the slight vertical offset between the mesial and distal lophs; **E,** stereo radicular view. The red dot is beneath the metastyle.

Diagnosis: Translated from Pickford (1987): "*small species in which the upper molar series is ca 41 mm long*"; from Lihoreau et al. (2004): "*Anthracotheriinae close to* Microbunodon minimum. *Premolar row longer than molar row. Geniohyoideus attach (sic) keeled and extended anteriorly in ventral view. Small parastyle. Symphysis ventral edge crested along entire length.* Microbunodon minimum *differs from* Mb. silistrensis *in equivalent length of premolar row relative to molar rows, in shorter absolute premolar row length, symphysis ventral edge being less crested with posterior flattening, its shorter and wider symphysis, its two large, well-marked genial fossae, its less transversally compressed mandible at level of c–p/1 diastema, and its strong parastyle.*"

Emendation to diagnosis: Lower third molars range in length from 25.4 to 27 mm. Differs from *Microbunodon minimum* (sensu Lihoreau et al. 2004) in that its upper second and third molars possess a well-developed postparaconule cristule that extends to the median transverse valley and contacts the premetaconule cristule, unlike *Mb. minimum* in which the postparaconule cristule stops short before reaching the median transverse valley. Upper molars possess a clear anterior accessory cusplet (protostyle) on the mesial cingulum and a prominent cusplet at the lingual end of the median transverse valley, unlike species of *Microbunodon* and *Microselenodon* in which the mesial cingulum has weak to absent swellings and the cusplet at the lingual end of the transverse valley is small or obsolete.

Lectotype: NHMUK PV OR 19041, right M3/ (Fig. 9).

Type locality: Caribari, Garo Hills, Bengal, India.

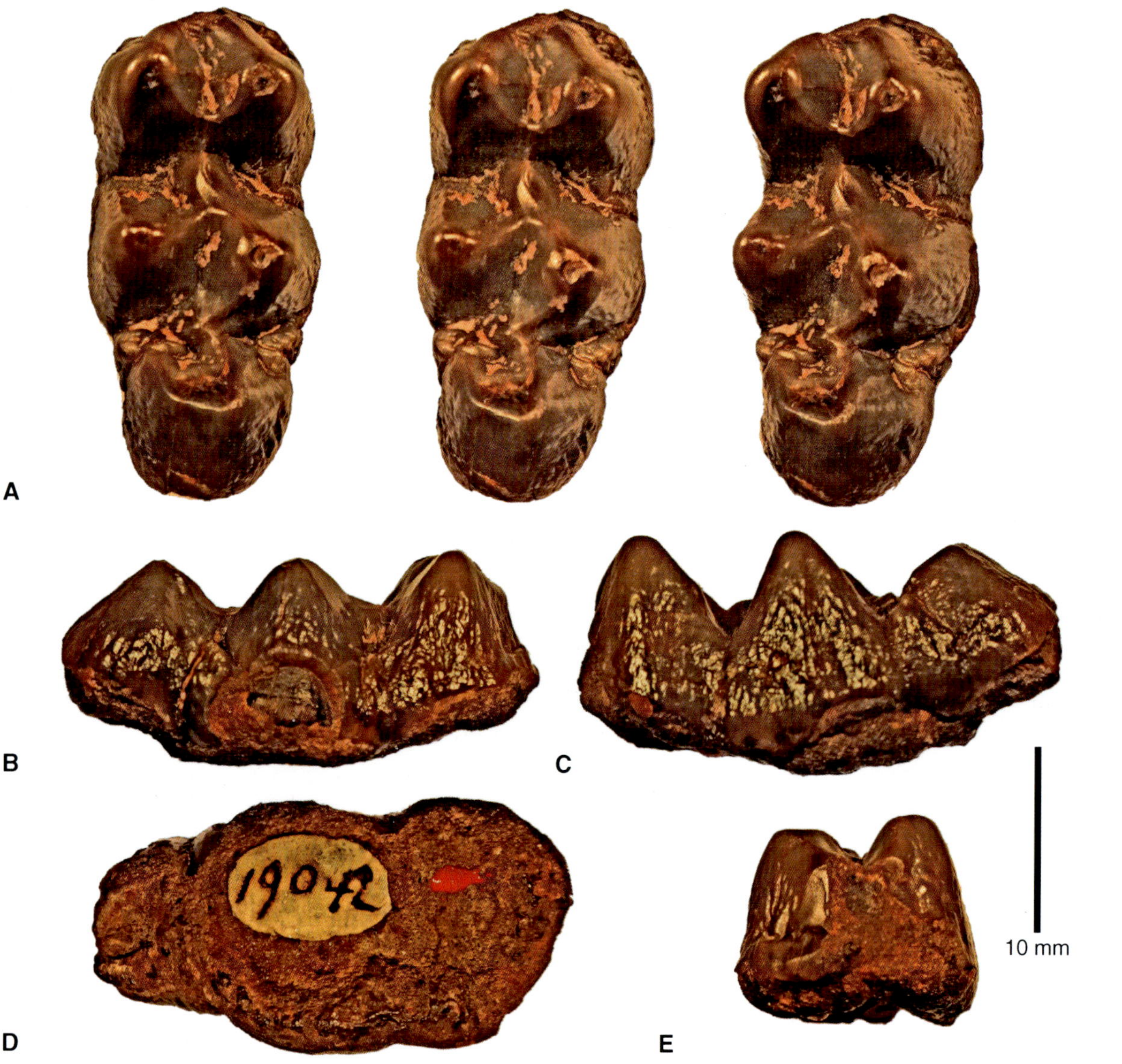

Figure 10. NHMUK PV OR 19042, right m/3 of *Garobunodon silistrensis* from the type locality of the species (Caribari, Bengal, India). **A,** stereo occlusal views; **B,** buccal view; **C,** lingual view; **D,** basal view to show the specimen number without the prefix 'M'; **E,** mesial view.

Distribution in Indo-Pakistan: Caribari, Sind, Dera Bugti, Chur Lando, Chinji, Ramnagar, Nagri (Dhok Mila)

Age: late Middle Miocene to basal Late Miocene (ca 12-9.5 Ma)

Hypodigm: (see Annex)

Descriptions

Most of the specimens herein attributed to *Garobunodon silistrensis* have already been described and some of them have been illustrated, so it is not necessary to redecribe all of them in detail. Some comments, however, will be made where appropriate.

The type specimen of the species is NHMUK PV OR 19041, a right M3/ in moderate wear, and lacking most of the roots (Fig. 9). The paraconule is large and deeply worn, and its postparaconule crista extends as far as the median transverse valley where it joins the premetaconule crista. The distal surface shows no signs of an interstitial contact facet, and in buccal view, a small vertical offset affects the distal loph and the distal roots are not overhung by the crown. These indications accord with the determination of its meristic position as M3/, rather than as M2/ or M1/. The enamel is coarsely but not heavily wrinkled.

NHMUK PV OR 19042 is a lightly worn right m/3 from Caribari, Bengal, with small dentine exposures at the apices of the five main cusps (Fig. 10). The preprotocristid and prehypocristid terminate close to the midline of the crown and the hypoconulid is centrally positioned. The metaconid has three short but inflated cristids, the postcristid extending across towards the postcristid of the protoconid, thereby closing off the distal part of the mesial fovea and there is a low relief mesiolingual metacristid. On the distolingual side of the metaconid there is a low-relief distolingual metacristid. The preprotocristid is long and swollen, and terminates anteriorly above the mesial cingulum in the midline of the tooth. There is no distobuccal protocristid. The prehypocristid is long and swells mesially such that it blocks off the median trans-

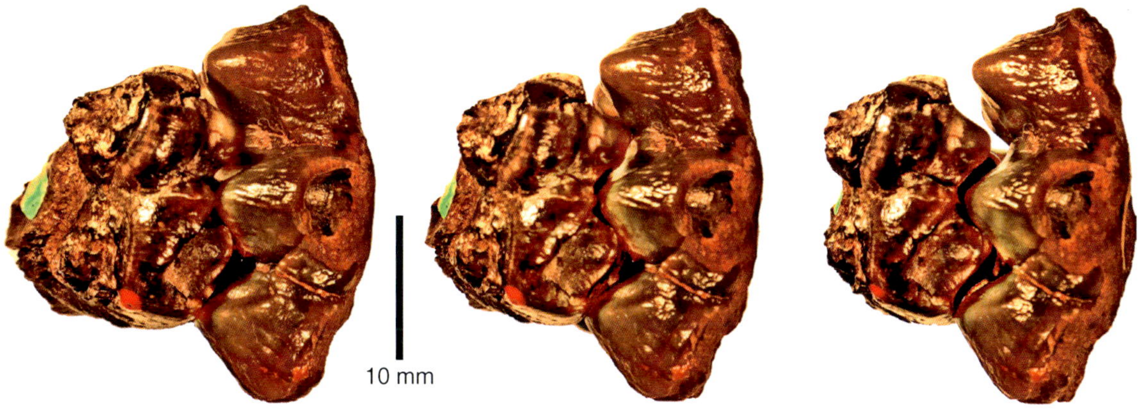

Figure 11. Stereo buccal views of the occluded right upper and lower third molars of *Garobunodon silistrensis* from Caribari, Bengal (NHMUK PV OR 19041: M3/ and PV OR 19042: m/3) to show the compatibility in dimensions and morphology of the two teeth (mesial to top of page).

verse valley, almost forming a median accessory cusplet. The endohypocristid is absent, but the posthypocristid is long and terminates in the midline of the tooth where it is joined by the prehypoconulid cristulid, the combined cristids blocking off the posterior transverse valley, but leaving the distal fovea open on its lingual side. The mesiolingual entocristid and postentocristid are short and weakly expressed, but the preentocristid is inflated and extends obliquely anteriorly towards the prehypocristid which it joins midway along its length, thereby separating the lingual part of the mesial transverse valley from the distal fovea and distal transverse valley. The enamel is coarsely, but not heavily wrinkled,

In order to test the possibility that the upper and lower third molars from Caribari belong to the same species, they were occluded together (Fig. 11). The fit is not perfect, suggesting that the two teeth could belong to different individuals, but it is close enough to remove any doubt that they belong to a single species. This demonstration is important in that it reveals what the lower molars of the species *Garobunodon silistrensis* would look like on the basis of a specimen from the type locality.

The upper molar row of *Garobunodon silistrensis* is represented by a maxilla from Chur Lando, first described by Forster-Cooper (1924, fig. 15c). A remarkable feature of the specimen (NHMUK PV M 12710, Fig. 12) is that

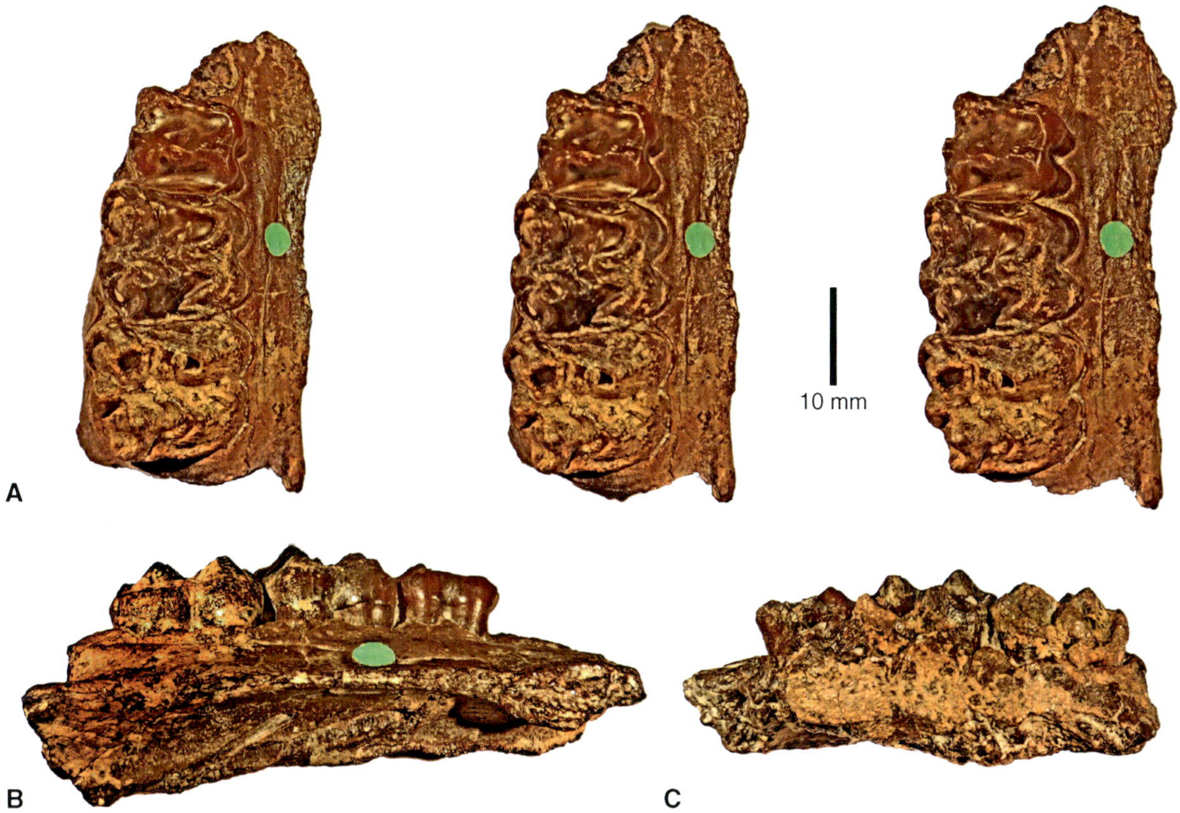

Figure 12. NHMUK PV M 12710, right maxilla with M1/-M3/ of *Garobunodon silistrensis* from Chur Lando, Pakistan. **A,** stereo occlusal views; **B,** lingual view; **C,** buccal view.

Figure 13. AMNH FM 94625, right M2/ of *Garobunodon silistrensis* from Ramnagar, Jammu & Kashmir, India. **A,** buccal view; **B,** occlusal view, C) distal view.

the M1/ is deeply worn to the stage that almost no occlusal morphology remains to study, yet the M3/ is only moderately worn with dentine lakes exposed at the apices of the protocone and paracone, but not on the metacone and metaconule.

A diversity of upper molars of *Garobunodon silistrensis* has been collected at Ramnagar (Jammu & Kashmir, India). The specimens were illustrated on Morphosource by Patel et al. (2022) and are herein illustrated in stereo occlusal view for convenience, and because some of them are in excellent condition, showing the cusp morphology exceptionally clearly (Figs 14, 15). The overall morphology of the crowns is close to that of the type specimen from Caribari. Notable features of the crowns are the pinched off mesostyle which forms a separate cusplet at the buccal end of the median transverse valley, and weak parastyle and metastyle that are not at all loop-like, thereby contrasting strongly with those of bothriodont anthracotheres. The paraconule is large and in the M3/s its postcrista extends across the median transverse valley to join the premetaconule crista. The protocone has a short precrista that ends close to the paraconule, and a stronger postcrista that descends into the median transverse valey. The lingual side of the protocone is swollen, especially in the posterior molars and it has a moderately developed lingual precrista. The pre- and postmetaconule cristae are strongly developed, the former one blocking the median transverse valley. The precrista and postcrista of the paracone and metacone are oriented almost in a straight line parallel to the long axis of the toothrow. The 'barrels' on the buccal surfaces of the paracone and metacone are not as clearly expressed as they are in bothriodont upper molars.

The upper molars possess a well-developed mesial cingulum that extends partly round the lingual side of the protocone where it diminishes in width before increasing again at the lingual end of the median transverse valley, where it forms a distinct basal pillar. The mesial cingulum also has a distinct accessory cusplet (protostyle) developed opposite the groove between the protocone and paraconule. There is a cingulum on the lingual and distal sides of the metaconule, that fades out before reaching the end of the postmetaconule crista. The buccal cingulum is weakly developed on the buccal side of the paracone, but is more in evidence on its mesial end. The metacone, in contrast, usually possesses a moderately developed buccal cingulum, in some specimens joining a clear posterior crest on the mesostyle.

The lower third molar of *Garobunodon silistrensis* from Ramnagar (WIHG FR 17/1) is closely similar in morphology to the specimen from Caribari, but it is at a slightly more advanced stage of wear (Fig. 15). The wear facet on the apex of the protoconid extends onto the pre- and postprotocristids, forming a crescentic dentine lake, which is confluent with the small dentine exposure on the postmetacristid.

Discussion

The recognition of the correct catalogue number of the right m/3, NHMUK PV OR 19042 (given as M19442 in Lihoreau et al. 2004) fundamentally modifies many previous ideas about the affinities and dimensions of *Garobunodon silistrensis*. The tooth is from Caribari, the same locality as the type specimen, and it is compatible with it in morphology and dimensions. One outcome is that the species *Microbunodon milaensis* is a junior synonym of *Garobunodon silistrensis*. Another conclusion is that *Garobunodon silistrensis* is the largest of the known species of *Garobunodon* from the Indian Subcontinent, and that *Garobunodon punjabiensis*, long thought to be bigger than *Garobunodon silistrensis*, is in fact smaller than it. The latter notion became established when Lydekker (1883, pl. 24, fig. 1, 1a) compared the holotype mandible fragment of *Anthracotherium punjabiense* with a lower jaw of a tragulid (*Dorcabune* Pilgrim, 1910a) that he misattributed to *Anthracotherium silistrense* (see Pilgrim, 1912), an erroneous attribution that influenced subsequent researches for almost a century and a half.

The present paper refines the hypodigm of *Garobunodon silistrensis* (Annex) and the reorganised data suggest that the species survived in the Indian Subcontinent from the late Middle Miocene to the basal Late Miocene (ca 12-9.5 Ma).

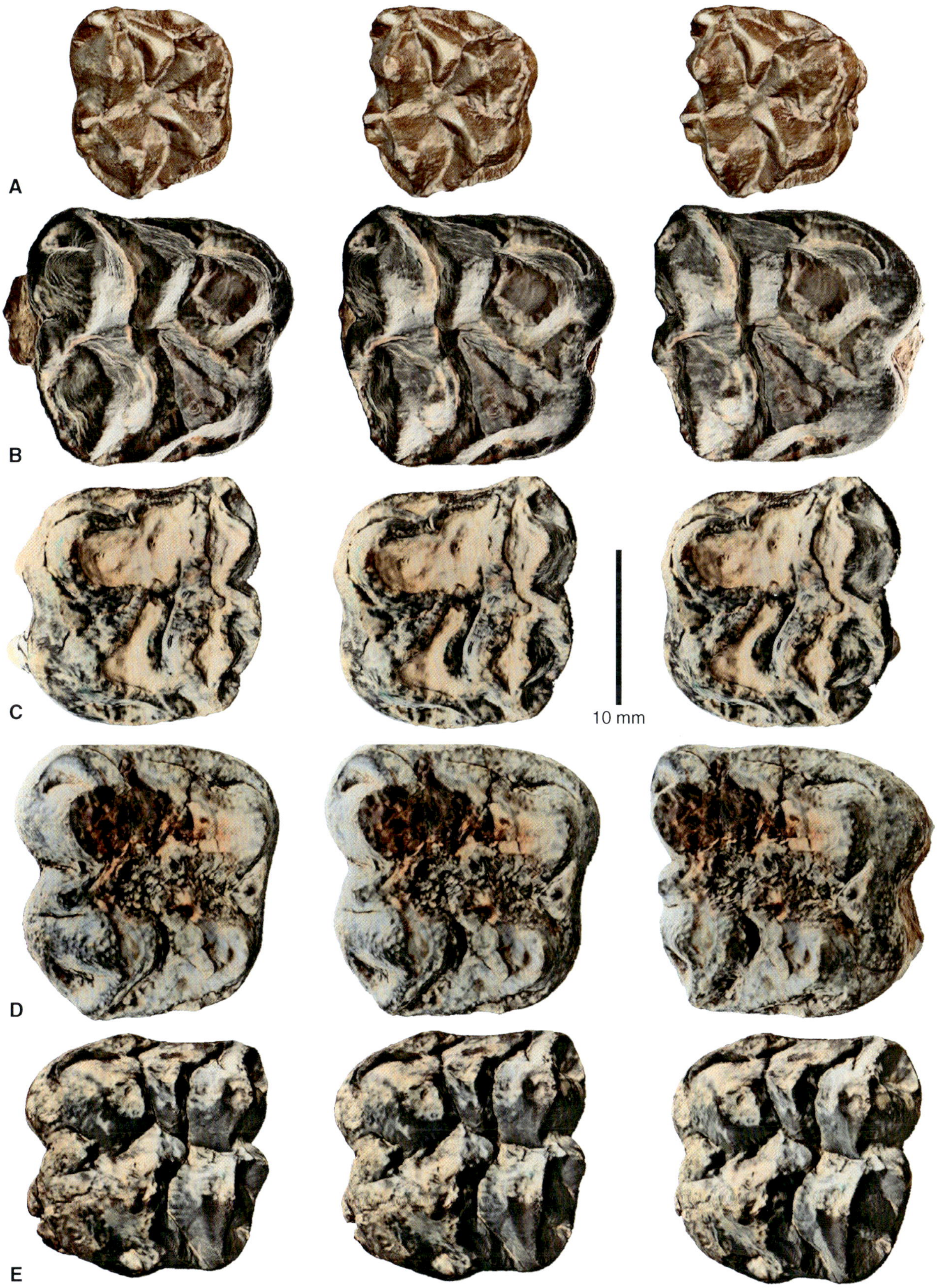

Figure 14. Stereo occlusal views of upper molars of *Garobunodon silistrensis* from Ramnagar (Jammu & Kashmir, India). **A,** WIHG RF 7, right M1/; **B,** YPM 20098, right M3/; **C,** WIHG R 23/1, left M3/; **D,** WIHG FR 21/L, right M3/; **E,** WIHG FR 11/1, left M3/ (base images extracted from Morphosource, Patel et al. 2022).

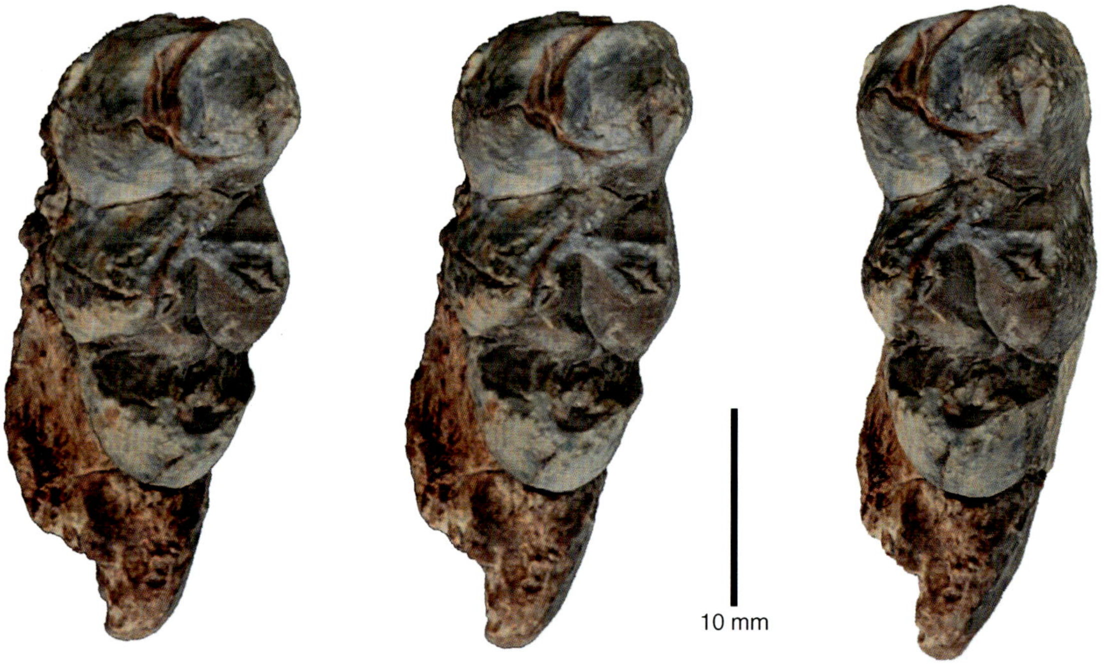

Figure 15. Stereo occlusal views of WIHG FR 17/1, left m/3 of *Garobunodon silistrensis* from Ramnagar, Jammu & Kashmir, India (base images extracted from Morphosource, Patel et al. 2022).

Species *Garobunodon punjabiensis* (Lydekker, 1877a)

Synonymy

v*1877a *Anthracotherium punjabiense* sp. nov. Lydekker, p. 78
1908 *Anthracotherium mus* sp. nov. Pilgrim, p. 150
1910b *A.* (*Microselenodon*) *mus* (Pilgrim); Pilgrim, p. 201
v1912 *Microselenodon silistrensis* (Pentland); Pilgrim, figs 2, 2a
v1912 *Microselenodon mus* (Pilgrim); Pilgrim, pp. 46-48, pl. 14, figs 3, 3a, 3b
1913 *Microbunodon mus* (Pilgrim); Pilgrim, p. 317
1935 *Anthracotherium punjabiense* Lydekker; Colbert, pp. 267-272, fig. 120
1976 *Anthracotherium punjabiense* Lydekker; Cabard, pp. 119-120
1976 *Anthracotherium mus* Pilgrim; Cabard, p. 120
1984 '*Anthracotherium*' *punjabiense* Lydekker; Raza et al. pp. 30-32
1984 *Anthracotherium punjabiense* Lydekker; Raza & Meyer, p. 51
1984 *Anthracotherium exiguum* Forster-Cooper; Raza & Meyer, p. 51
pv2004 *Microbunodon silistrensis* (Pentland, 1828); Lihoreau et al. 2004, pp. 103-107, fig. 5, tab. 3
pv2004 *Microbunodon milaensis* Lihoreau, Blondel, Barry & Brunet, pp. 107-109, fig. 6, tab. 3
p2013 *Microbunodon silistrensis* (Pentland, 1828); Khan et al. p. 32, tab. 9, fig. 22a (left D4/-M1), (non fig 22b, rear two lophids of left d/4, *Conohyus* sp. or *Kachchhchoerus* sp; non fig. 22d, left d/4-m/1 of *Merycopotamus nanus*)

Diagnosis: Medium-sized species of *Garobunodon*, smaller than *Garobunodon silistrensis* and larger than *Garobunodon exiguus*. Lower third molars range in length from 21.0-23.4 mm compared to 24.5-27 mm in *Garobunodon silistrensis* and 18.3-20.4 mm in *Garobunodon exiguus*. Well-developed postparaconule cristule extending to the median transverse valley and contacting the premetaconule cristule.

Holotype: GSI B 104, right mandible with m/2-m/3 (Fig. 16).

Type locality: Salt Range (?Hasnot), Pakistan

Distribution in Indo-Pakistan: Salt Range (?Hasnot), Chinji, Kanatti, Khumbi, Chur Lando, Dera Bugti, Sind

Age: late Early Miocene to basal Middle Miocene (ca 17-12 Ma)

Hypodigm: (see Annex)

Descriptions

Most of the available specimens of *Garobunodon punjabiensis* have been described and illustrated on a number of occasions (Lydekker, 1877a, 1878; Pilgrim, 1912; Forster-Cooper, 1924; Pickford, 1987; Lihoreau et al. 2004) but they have often been attributed to *Microbunodon silistrensis* (see synonymy list above). It is not necessary to redescribe the material in detail, but herein some of the more important specimens are illustrated. One of the specimens attributed to this species by Forster-Cooper

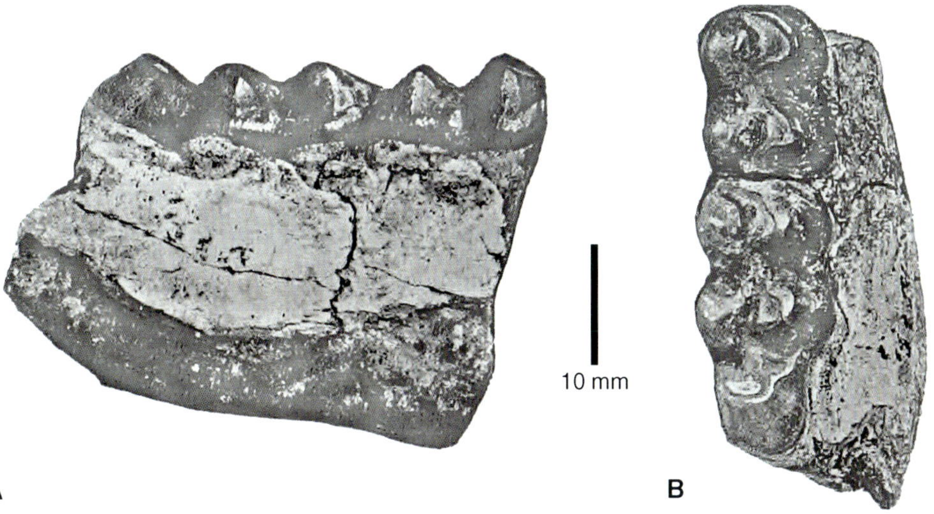

Figure 16: GSI B 104, right mandible fragment with m/2-m/3, the holotype of *Garobunodon punjabiensis* (LYDEKKER, 1878) from the Lower Siwaliks of the Salt Range (?Hasnot) Potwar Plateau, Pakistan. **A;** lateral view, **B;** occlusal view (images modified from Pilgrim, 1912, pl. XIV, figs 2, 2a).

(1924, fig. 17) belongs in fact to *Garobunodon silistrensis*. In his plate VII, fig. 3, the author attributed the same tooth to *Anthracotherium punjabiense?*, which suggests that he had some doubts about its species affinities. Its dimensions and morphology are close to those of the Caribari tooth (NHMUK PV OR 19042) that belongs to *Garobunodon silistrensis*, and it is appreciably larger than the type specimen (GSI B 104) of *Garobunodon punjabiensis* (Fig. 16).

The wear gradient in the m/2 and m/3 of the left mandible fragment (NHMUK PV M 12708, Fig. 17) accords with the style of wear in the maxillary molars of NHMUK PV M 12710 from Chur Lando (Fig. 12). The m/2 is deeply worn to the stage where there is a single large dentine

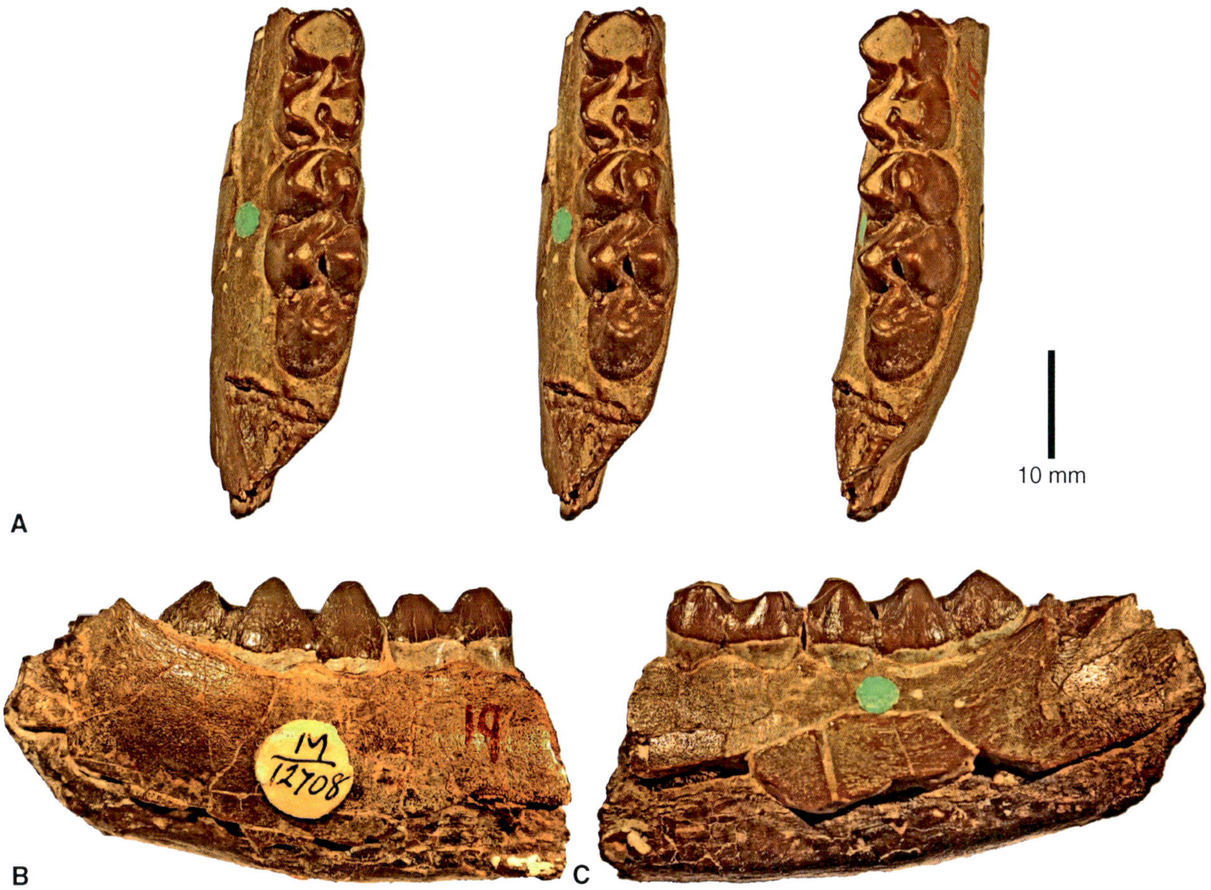

Figure 17. NHMUK PV M 12708, left mandible with m/2-m/3 of *Garobunodon punjabiensis* from Dera Bugti, Pakistan. **A,** stereo occlusal views; **B,** lingual view; **C,** buccal view. This specimen was included in *A. silistrense* by Forster-Cooper (1924, fig 16).

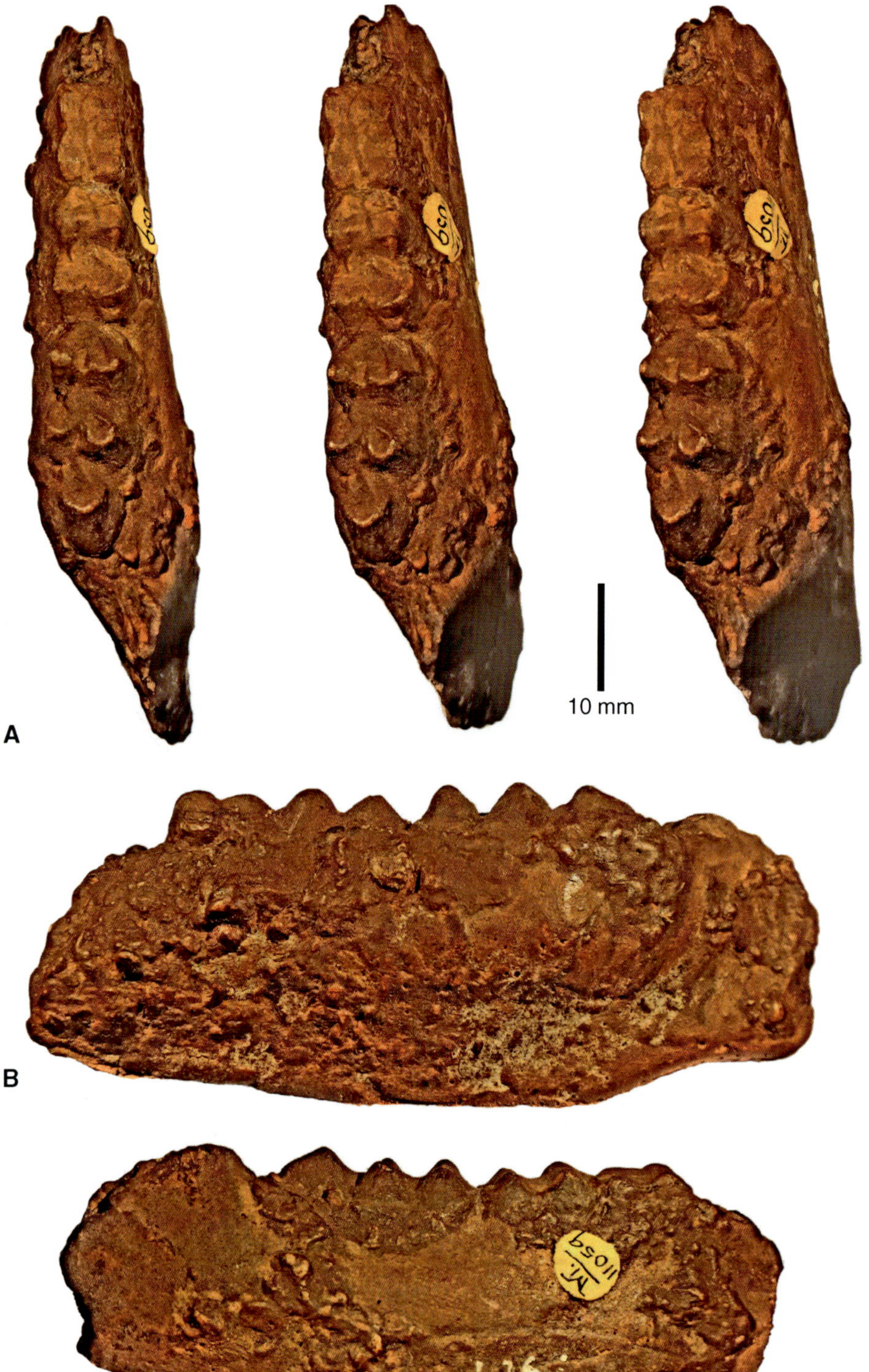

Figure 18. NHMUK PV M 11059, cast of GSI B466, right mandible with m/1-m/3 of *Garobunodon punjabiensis* (= holotype of *Microselenodon mus* (PILGRIM, 1908, 1912) from Khumbi, Pakistan. **A,** stereo occlusal views; **B,** lingual view; **C,** buccal view.

lake on the mesial lophid, whereas the m/3 is much less worn, with the dentine lakes of the cusps in each lophid separated from each other. The exaggerated wear gradient in the posterior teeth of *Garobunodon* sets it apart from the much less marked wear gradient that occurs in bothriodontine molar rows.

GSI B466, the right mandible with m/1-m/3 of *Garobunodon punjabiensis* (holotype of *Microselenodon mus* (PILGRIM, 1912)) from Khumbi (spelled Kumbhi by Pilgrim, 1912), shows the exaggerated wear gradient expressed in the molar row that is characteristic of species of *Garobunodon* (Fig. 18). The m/1 is worn down to the stage where there is but a single dentine lake encompassing the mesial and distal lophids, the ensemble being enclosed by a thin vertical wall of enamel, and in the m/2 the dentine lakes on the mesial and distal lophids are almost joined together, yet in the m/3 the dentine lakes at the apices of the five main cusps are still separated from each other.

Discussion

The demonstration that *Garobunodon silistrensis* is larger than *Garobunodon punjabiensis* rather than smaller than it, as thought by most authors since Lydekker (1883, pl. 24, fig. 1, 1a) misidentified a mandible of the large tragulid *Dorcabune* as the jaw of *Anthracotherium silistrense* (Pilgrim, 1912) means that many of the Indo-Pakistan specimens previously attributed to species on the basis of their dimensions, need to be reassigned (see Annex).

Garobunodon exiguus (Forster-Cooper, 1924)

Synonymy

v*1924 *Anthracotherium exiguum* sp. nov. Forster-Cooper, p. 21, figs 18a, 18b, 18c

pv1987 *Anthracotherium silistrense* Pentland; Pickford, pp. 313-316.

p2004 *Microbunodon silistrensis* Pentland; Lihoreau et al. table 3.

Diagnosis: Small species of *Garobunodon*, smaller than *Garobunodon punjabiensis*. Lower third molars range in length from 18.3-20.4 mm, compared to 21.0-23.4 mm in *Garobunodon punjabiensis* and 24.5-27 mm in *Garobunodon silistrensis* (Fig. 24).

Lectotype: NHMUK PV M 12055, right maxilla fragment with worn M2/-M3/ (Forster-Cooper, 1924, fig. 18b).

Paralectotype: NHMUK PV M 12054, left maxilla with heavily worn M1/-M2/ (Forster-Cooper, 1924 fig. 18c).

Type locality: Dera Bugti, Pakistan

Distribution in Indo-Pakistan: Dera Bugti, Chur Lando, (?)Chinji, (?)Sind.

Age: Early Miocene (ca 18.5-16 Ma).

Hypodigm: (see Annex)

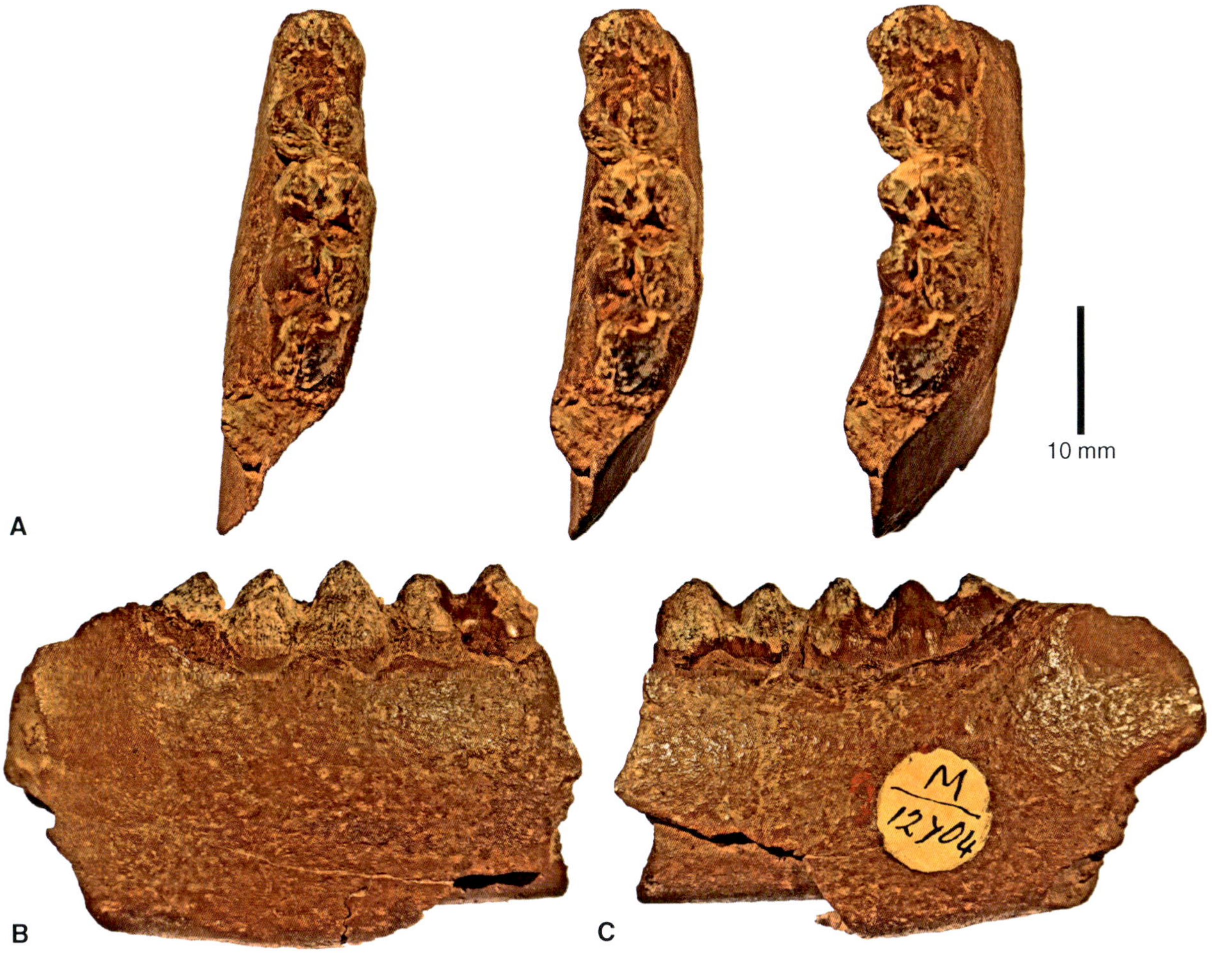

Figure 19. NHMUK PV M 12704, left mandible with m/2-m/3 of *Garobunodon exiguus* from Chur Lando, Pakistan. **A,** stereo occlusal views; **B,** lingual view; **C,** buccal view.

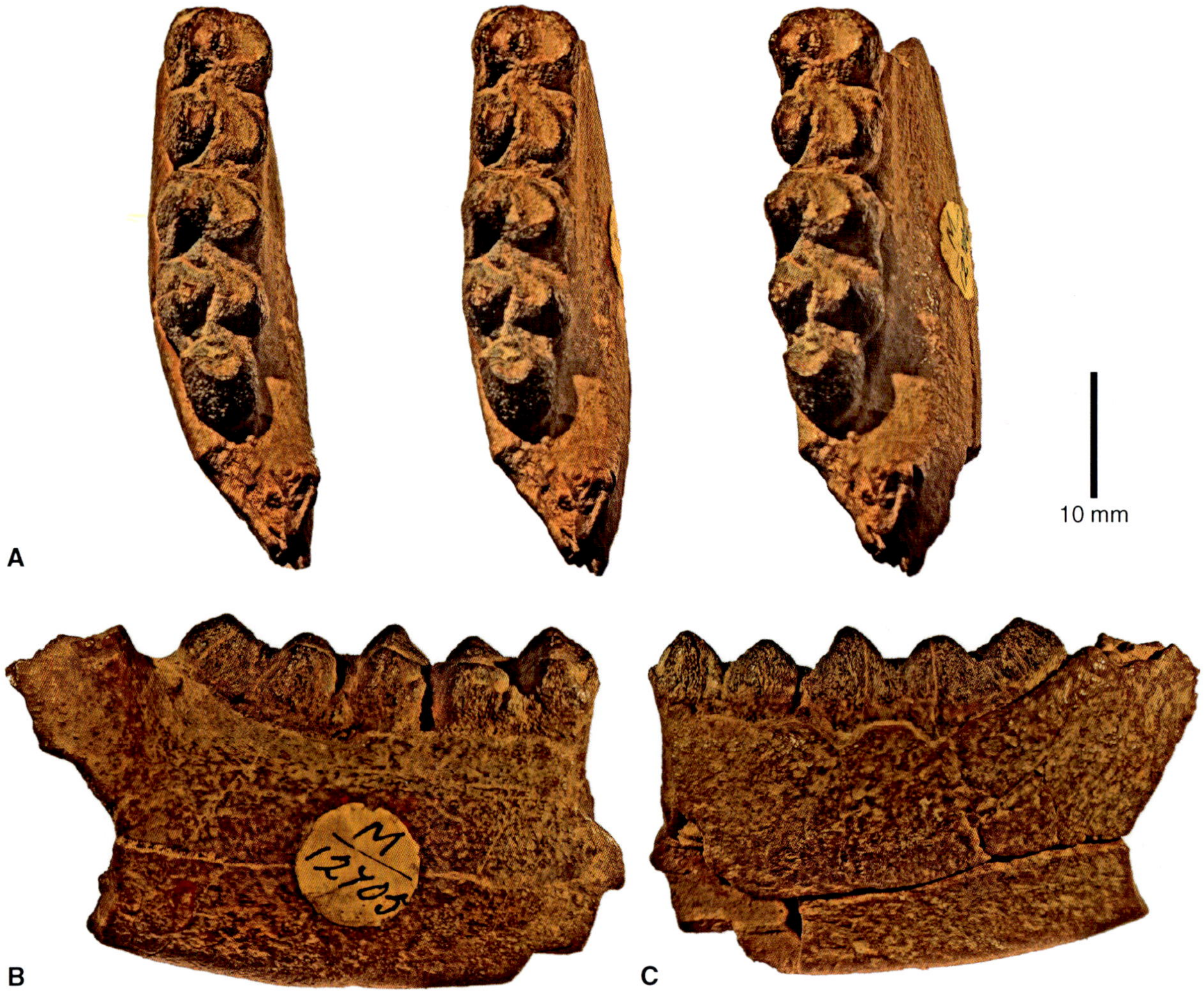

Figure 20. NHMUK PV M 12705, right mandible with m/2–m/3 of *Garobunodon exiguus* from Chur Lando, Pakistan. **A,** stereo occlusal views; **B,** buccal view; **C,** lingual view.

Description

The maxillae and upper teeth of *Anthracotherium exiguum* were illustrated by Forster-Cooper (1924, figs 18a, 18b, 18c). The teeth in all three specimens are so worn that little can be discerned about their morphology. The M3/ in the lectotype is heavily worn, but it is possible to see that it was probably pentacuspidate. The M1/ in the paralectotype is substantially smaller than the M2/, a feature also present in *Microselenodon minimum* and *Microbunodon laharpei* from Europe. The main criterion employed by Forster-Cooper (1924) when erecting the new species was the observation that the teeth are smaller than those of *Anthracotherium punjabiense* and *Anthracotherium silistrense*, although the author pointed out that the upper molars of his new species had a smaller metastyle than that of *A. silistrense*, and that its lower molars were "*also smaller than in* A. silistrense *and differ also in the entaconid (sic) which is triangular through the absence of ridge No.* 3". The four mandibles allotted to the species by Forster-Cooper (1924, p. 21) were not illustrated, but they are well-preserved (Figs 19–21) and are more informative than the maxillary fragments.

All the upper molars are worn to the stage that the original extent of the metastyle is not discernible, so this supposed difference between the species is not soundly based, and the only supposed morphological difference between the lower molars of these species (presence or absence of ridge n° 3 in the entoconid) does not hold up to scrutiny, because, like *Garobunodon exiguus*, neither *Garobunodon silistrensis* nor *Garobunodon punjabiensis* possesses such a ridge (posterointernal cristid).

Examination of three of the mandibles attributed to *Anthracotherium exiguum* (NHMUK PV M 12704–12706, Figs 19–21) shows that the molars, which are relatively well preserved, are indeed similar in most morphological details to their counterparts in *Garobunodon silistrensis* and *Garobunodon punjabiensis* and the specimens are herein included in this genus rather than in *Anthracotherium*. There are minor differences in the strength of the postmetacristid and the preentocristid, but these are probably examples of individual variation. Differences in wear stages and small patches of matrix attached to the teeth also affect proper appreciation of their morphology.

Discussion

As concerns *Anthracotherium exiguum*, the specimens attributed to it by Forster-Cooper (1924, figs 18a, 18b, 18c: NHMUK PV M 12053–12055) are smaller than material of *Garobunodon silistrensis* and *Garobunodon punjabi-*

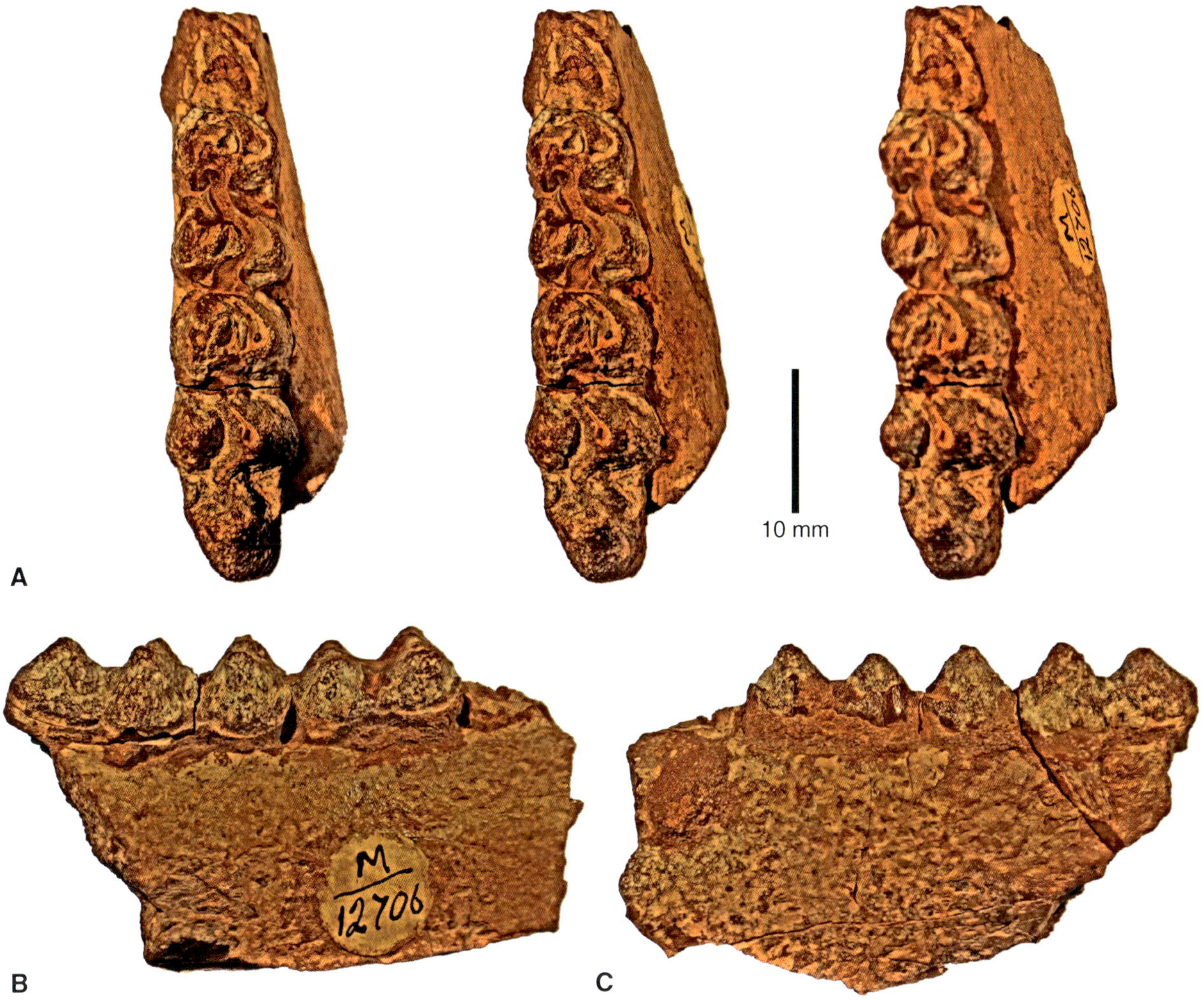

Figure 21. NHMUK PV M 12706, right mandible with m/2–m/3 and roots of m/1 of *Garobunodon exiguus* from Chur Lando, Pakistan. **A,** stereo occlusal views; **B,** buccal view; **C,** lingual view.

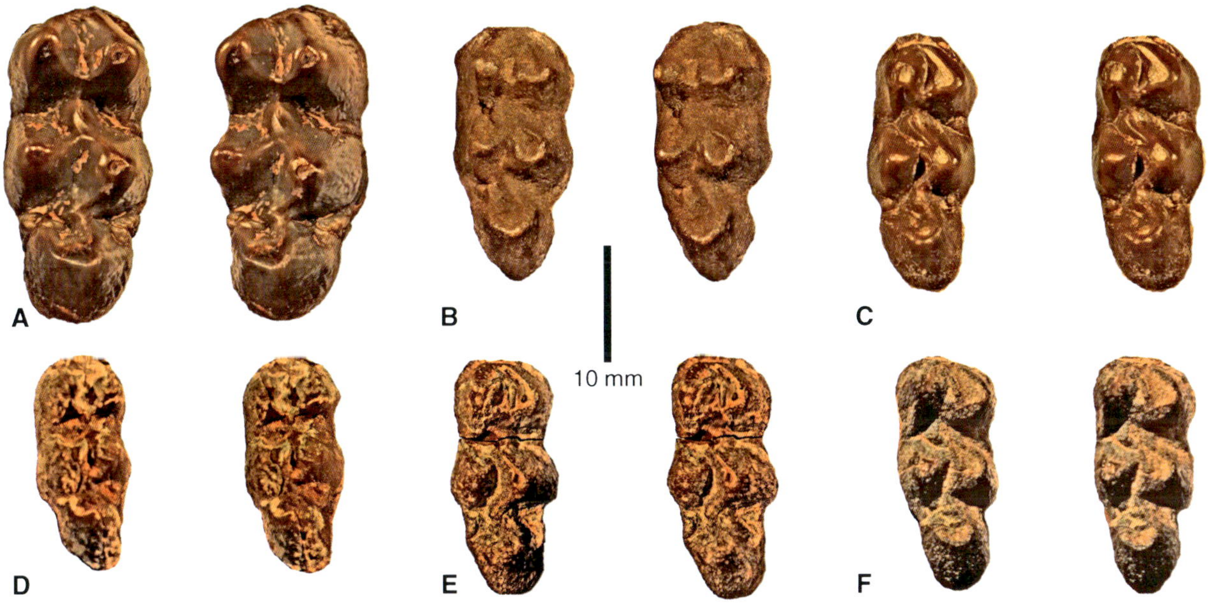

Figure 22. Stereo occlusal views of lower right third molars of *Garobunodon* from Indo-Pakistan. **A,** *Garobunodon silistrensis*, NHMUK PV OR 19042; **B–C,** *Garobunodon punjabiensis* (PV M 11059 and PV M 12708 (image reversed)); **D–F,** *Garobunodon exiguus* (PV M 12704 (image reversed), PV M 12706 and PV M 12705).

ensis, as are the lower molars in three of the mandibles (NHMUK PV M 12704–12706) that he mentioned but did not describe. The teeth in the palate are poorly preserved, as are those in the other two maxilla fragments that he illustrated. Herein the maxilla fragment with the least worn molars (NHMUK PV M 12055) is nominated as the lectotype, as it is the most informative of the three specimens illustrated by Forster-Cooper (1924). The lower molars in contrast are better preserved, and they reveal that the specimens are close in morphology to those of *Garobunodon silistrensis* and *Garobunodon punjabiensis*, but differ from these two taxa by their more diminutive dimensions, as was noted by Forster-Cooper (1924) (Fig. 22).

The species *Microbunodon sminthos* FORSTER-COOPER, 1913, that was substantially smaller than *Garobunodon exiguus*, was shown by Pickford (1987) to be based on a specimen of the suoid *Pecarichoerus* Colbert, 1933.

The presence of a marked wear gradient in the molar rows of the three species of *Garobunodon* is an additional similarity between them, suggesting that their diets were probably rather comparable.

Metric analyses

Historical background

There has been a longstanding misconception about the relative dimensions of the species of *Garobunodon* from Indo-Pakistan (in the literature specimens have been variably included in *Anthracotherium* CUVIER, 1822; *Choeromeryx* POMEL, 1848; *Microbunodon* DEPÉRET, 1908; and *Microselenodon* DEPÉRET, 1908). Until 2013, authors generally considered that *'Anthracotherium' silistrense* was smaller than *'Anthracotherium' punjabiense*, yet the holotype M3/ of *Garobunodon silistrensis* is the largest specimen of the genus recovered from the subcontinent. Part of the trouble was due to the fact that, in collections, upper molars are relatively scarce compared to lower molars, such that comparisons tended to marginalise the type and only known specimen of *Garobunodon silistrensis* from Caribari, and to focus comparisons on the lower third molars which were more common in the collections.

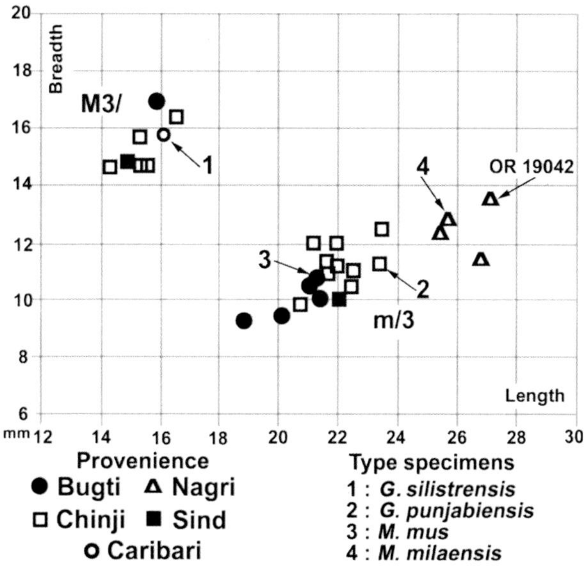

Figure 23. Bivariate (length-breadth) plots of upper and lower third molars of *Garobunodon* from Indo-Pakistan. Data is from Lihoreau et al. (2004, fig. 4, *M. Microbunodon*). The numbered points are the type specimens arranged in the order in which they were erected. Also included is the m/3 from Caribari (NHMUK PV OR 19042, not M19442 as given by the authors, nor was it from Nagri, but from Caribari).

However, the long forgotten lower third molar from Caribari (NHMUK PV OR 19042 = M19442 in Lihoreau et al. 2004) is the largest of its kind from the subcontinent, thereby reinforcing the conclusion that *Garobunodon silistrensis* is the largest species in the genus. The Caribari m/3 is slightly longer and broader than the m/3 in the holotype of *Mincrobunodon milaensis*. The latter name was erected by Lihoreau et al. (2004) mainly on the basis of its perceived greater dimensions compared to those of *Garobunodon silistrensis* as revealed by the first line of the diagnosis of the species *Micobunodon milaensis* ('*m/3 larger than in* Microbunodon silistrensis *and* Microbunodon minimum').

When Lydekker (1877a) erected the species *Anthracotherium punjabiense*, he did not compare it to the Caribari species, partly because he included the latter in the genus *Choeromeryx*, as *C. silistrensis*, and partly because lower molars of *C. silistrensis* had not been published, so he was unable to make comparisons. He provided measurements of the teeth of *A. punjabiense* in inches (converted to mm in Annex). At the time, he seems to have been unaware of the existence of NHMUK PV OR 19042, but in his subsequently published catalogue of Siwalik Mammals (Lydekker, 1885a) he listed this fossil under *Anthracotherium* sp. c, and wrote that it was from the Pliocene of the Siwalik Hills, India, and that it was larger than the m/3 in the cast M 1543 that he identified as *A. silistrense* (in reality a cast of a mandible of the tragulid *Dorcabune* PILGRIM, 1910a: see Pilgrim, 1912). Thus, Lydekker's (1877a) observation that *A. punjabiense* was larger than *A. silistrense* was based on comparison with a tragulid mandible erroneously attributed to the latter species. Furthermore, the handwritten catalogue in the NHMUK (Old Register) lists OR 19042 as *Choeromeryx* (*Anthracotherium*) *silistrense* and its locality as Caribari, Cooch-Behar, NE border of Bengal, thereby differing from the subsequently published 1885 catalogue in which it is listed as a Pliocene fossil from the Siwalik Hills. Pilgrim (1912) wrote that the specimen was from Assam.

The holotype of *Garobunodon punjabiensis* (LYDEKKER, 1877a) was included by Pilgrim (1912, p. 47, pl. 14, fig. 2) in *Microselenodon silistrensis*. Likewise, Lihoreau et al. (2004, table 3) concluded that *Microbunodon punjabiensis* was a junior synonym of *Microbunodon silistrensis*.

Lihoreau et al. (2004, fig. 4) presented bivariate plots of the upper and lower third molars of *Microbunodon* from

Europe and Indo-Pakistan. Their results differ slightly from the present analysis partly because the axes of their graphics were the mesiodistal length against the buccolingual breadth of the second loph(id), whereas in the present study the y-axis is the breadth of the mesial loph(id). The intervals on the x-axis and y-axis were different in their two graphics (for m/3 and M3/) which renders comparisons difficult, so their data on the Indo-Pakistan fossils are herein presented in the same format as that of the present analyses so that readers can readily make comparisons (Fig. 23).

Metric analysis present study

The analyses of the dental dimensions of small bunodont anthracotheres from the Indian subcontinent (Figs 24, 25) indicate that the holotype of *Garobunodon punjabiensis* from the '*Lower Siwaliks of the Salt Range area*' (Pilgrim, 1912) is smaller than the species represented at Caribari (*Garobunodon silistrensis*). The data support the conclusion that *Garobunodon silistrensis* is the senior name for the largest of the three taxa diagnosed from Indo-Pakistan, and that *Garobunodon punjabiensis* is a distinct species that was somewhat smaller than it.

The quantity of premolars of bunodont anthracotheres from Indo-Pakistan is too restricted to be of statistical relevance. Herein the focus of the metric analysis is on the molars which indicate that there are three size groups in the Miocene deposits of Indo-Pakistan.

The samples of upper first and second molars are limited, and some of the teeth are deeply worn, to the extent that length and breadth measurements need to be estimated, which introduces a certain amount of uncertainty into the analyses (Fig. 24). However, the few available specimens are included in the study so as to complete the analyses. Despite the limited amount of data, the few anterior molars that are available reveal the same pattern of metric variation that is revealed by the more abundant third molars.

Morphologically, all the third molars are similar to each other, indicating that they belong within a single genus, *Garobunodon*, but bivariate plots of mesiodistal length against buccolingual breadth (Fig. 25) indicate that there are three size groups, suggestive of species distinctions. The specimens in the largest of the three groups are herein attributed to *Garobunodon silistrensis*, the medium-sized ones to *Garobunodon punjabiensis* and the smallest ones to *Garobunodon exiguus*, the upper dentition of which is still poorly known.

Previous authors frequently transposed specimens between different species, and new species such as *Anthracotherium (Microselenodon) mus* PILGRIM, 1908 (1912) and *Microbunodon milaensis* LIHOREAU et al. 2004, were erected largely on the basis of dimensions that the authors perceived to be different from those of the previously established species.

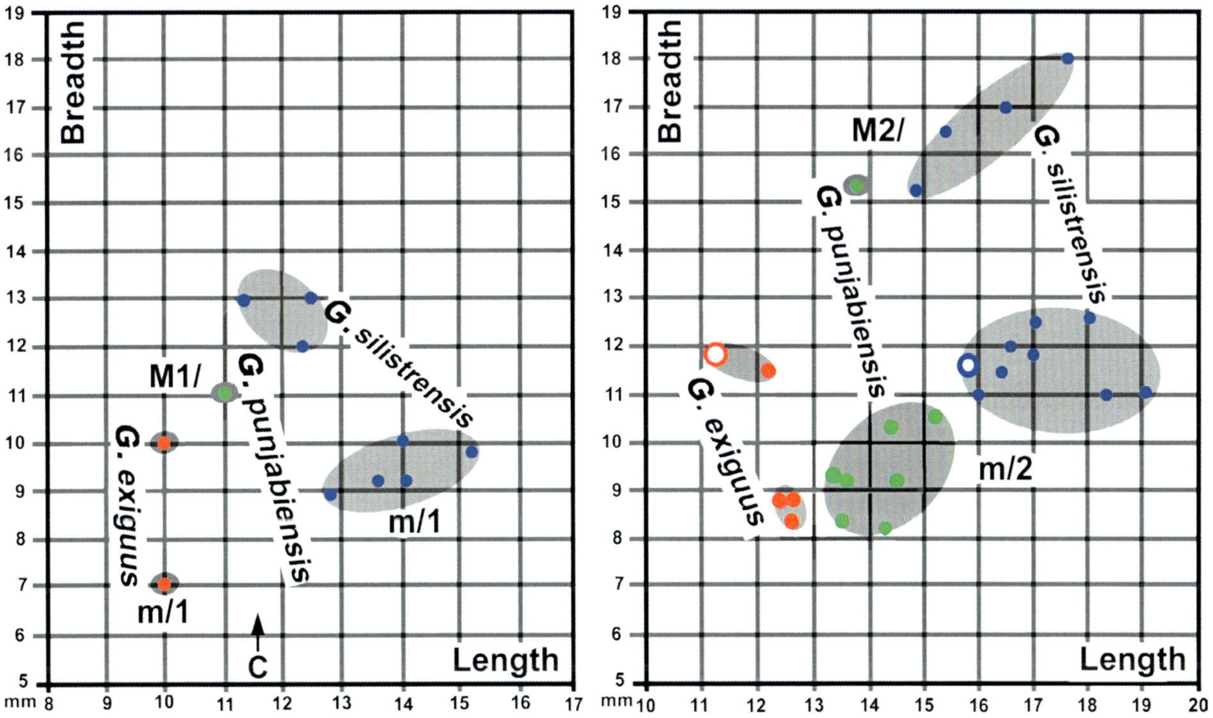

Figure 24. Bivariate (length-breadth) plots of upper and lower first (left) and second molars (right) of *Garobunodon* from Indo-Pakistan (blue dots, *Garobunodon silistrensis*; green dots, *Garobunodon punjabiensis*; red dots, *Garobunodon exiguus*). The large blue dot with a white center is the m/2 in the holotype of *Microbunodon milaensis* (YPM 10999 from Dhok Mila) and the large red dot with a white center is the M2/ in the lectotype of *Garobunodon exiguus* (measurements estimated). The m/1 labelled 'C' with an arrow is an m/1 from Chinji that lacks the breadth measurement (Lihoreau et al. 2004).

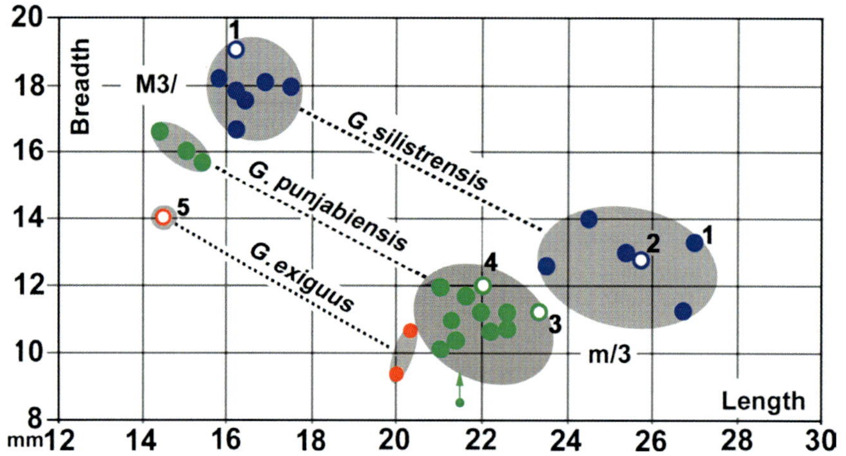

Figure 25. Bivariate (length-breadth) plots of upper and lower third molars of bunodont anthracotheres from the Miocene deposits of Indo-Pakistan. Dots with white centers represent type specimens: **1,** Caribari (*Garobunodon silistrensis*): **2,** Dhok Mila (*Microbunodon milaensis*); **3,** Hasnot(?) (*Garobunodon punjabiensis*); **4,** Khumbi (*Microselenodon mus*); **5,** Dera Bugti (*Garobunodon exiguus*: measurements of M3/ estimated). The m/3 numbered '1' is NHMUK PV OR 19042 from Caribari, that is close in size and morphology to the holotype of *Microbunodon milaensis* (number '2').

Biostratigraphic implications

In his discussion about an upper molar of '*Anthracotherium' silistrense* from Chenji (sic), Forster-Cooper (1924) hypothesised that "*This and the type specimen presumably both come from later zones than the Bugti deposits*". This biochronological suggestion is supported by the present analysis, and it accords with the trend frequently observed among fossil mammal lineages in which the body size of species increases with the passage of geological time (often called Cope's Rule (Cope 1885), but which was first explicitly articulated by Depéret (1907)).

Concerning the utility of anthracotheres for biostratigraphic studies, Matthew (1929, p. 453) commented that "*The phylogeny of this group and identification of genera and species have been so bedevilled by hasty and incomplete studies that they are hardly usable for correlation*". This sentiment is understandable when the convoluted history of their taxonomic interpretation has not been clarified, but it does not automatically render the anthracotheres worthless for throwing light on the passage of geological time. Correlations need to be made on the basis of the morphology and dimensions of the fossils rather than on the names that they have been given.

Forster-Cooper's (1924) prescient interpretation of the chronological positions of the so-called Bugti faunas relative to those from the Potwar Plateau (Chinji and Nagri 'zones') and Bengal, based on the small bunodont anthracotheres that he included in '*Anthracotherium*', appears to be valid (Pickford, 1988), and is supported by the present study, even though some of the specimens in the old collections have relatively imprecise locality data. The smallest of the three species of *Garobunodon* (*G. exiguus*) was collected from Dera Bugti and Chur Lando, probably from the Early Miocene part of the succession, estimated to be ca 18.5-16 Ma, the medium-sized species (*G. punjabiensis*) occurs predominantly in deposits of late Early Miocene and earlier Middle Miocene age (ca 17-12 Ma), whereas the largest species, *G. silistrensis*, occurs in deposits dating from the end of the Middle Miocene and into the base of the Late Miocene (12-9.5 Ma) (Fig. 8).

Localities such as Chinji, Sind, Dera Bugti and Chur Lando, that have yielded remains attributed to two or three of these anthracothere species, consist of thick sequences of strata spread over many square kms of exposure that likely encompass the stratigraphic ranges of the taxa involved. Thus their seeming 'association' at these 'sites' is likely to be an artefact of the collecting methods employed in the early days of research in the regions, rather than to the co-existence of the different species. In order to resolve the uncertainties, additional researches need to be undertaken in the future, during which detailed stratigraphic positions of fossils should be recorded.

Palaeodiet

Little has been written about the potential diet of the bunodont anthracotheres, but the presence of a marked wear gradient in the posterior teeth of all three species of *Garobunodon* from Indo-Pakistan indicates that their main source of food may have been abrasive and/or hard. The upper and lower first molars in the lineages are often worn to stubs before the third molars develop dentine lakes on the apices of their cusps. It is possible that their diet consisted not only of durable resources, but also that the food items may have been coated in sand or other abrasive substances, much like sea-weed on the shores of ancient seas, or tubers excavated from sandy soils and ingested without removal of the adherant soil.

The European species of *Microbunodon* (and *Microselenodon*) also show a strong wear gradient in the molar row (Hünermann, 1964; Lihoreau et al. 2004, fig. 2).

Phylogeny

Lihoreau et al. (2004, fig. 7) performed a cladistic analysis of the Anthracotheriinae, but the relationships between taxa require recalculation because at least one of the supposed apomorphic character states defined for *Microbunodon silistrense* (sic) does not occur in the species, to wit, the presence of a well-defined premetacristid in the m/3. Furthermore, their diagram of the m/3 of *Microbunodon milaensis* shows a strongly developed endoprotocristid, but none of the available specimens possesses this cristid, including their specimen NHMUK M19442 (= NHMUK PV OR 19042, the m/3 from, Caribari, the type locality of *Garobunodon silistrensis*). In addition the lack of character states for the M3/ of *Microbunodon milaensis* is problematic.

Nevertheless, it is clear that *Garobunodon* belongs to the subfamily Microbunodontinae Lihoreau & Ducrocq, 2007, and that it shares characters with other genera of this subfamily: *Microbunodon*, *Microselenodon*, *Anthracokeryx* Pilgrim & Cotter, 1916, *Nabotherium* Sileem, Sallam, Hewaidy, Miller & Gunnell, 2016, *Geniokeryx* Ducrocq, 2020, and *Oligokeryx* Ducrocq, Chaimanee, Yamme & Jaeger, 2025. As such *Garobunodon* brings to seven the quantity of genera known in the subfamily. If the Tuscan genus *Etruscotherium ribollaense* Pickford, 2021, is added, then the diversity would increase to eight genera.

Palaeobiogeography

Fossils attributed to Microbunodontinae have been recorded from many localities in Europe and Asia (Lihoreau et al. 2004) but thus far have been found in only one site in Africa (Fayum, Egypt; Sileem et al. 2016) which was close to the southern shores of the Palaeo-Tethys Ocean (Fig. 26).

Some lineages of small anthracotheres (bothriodontines such as *Sivameryx* and *Merycops*) that occur in Indo-Pakistan side-by-side with *Garobunodon*, have been found in Africa at sites such as Moghara (Egypt; Miller et al. 2014), Gebel Zelten (Libya; Pickford, 1991) and diverse sites in Kenya (Pickford, 1991).

It appears from this pattern of distribution, that the bothriodonts may have been adapted to the more tropical and sub-tropical latitudes of the Old World, and that Oligo-Miocene microbunodonts were adapted to more boreal, mid-latitudes.

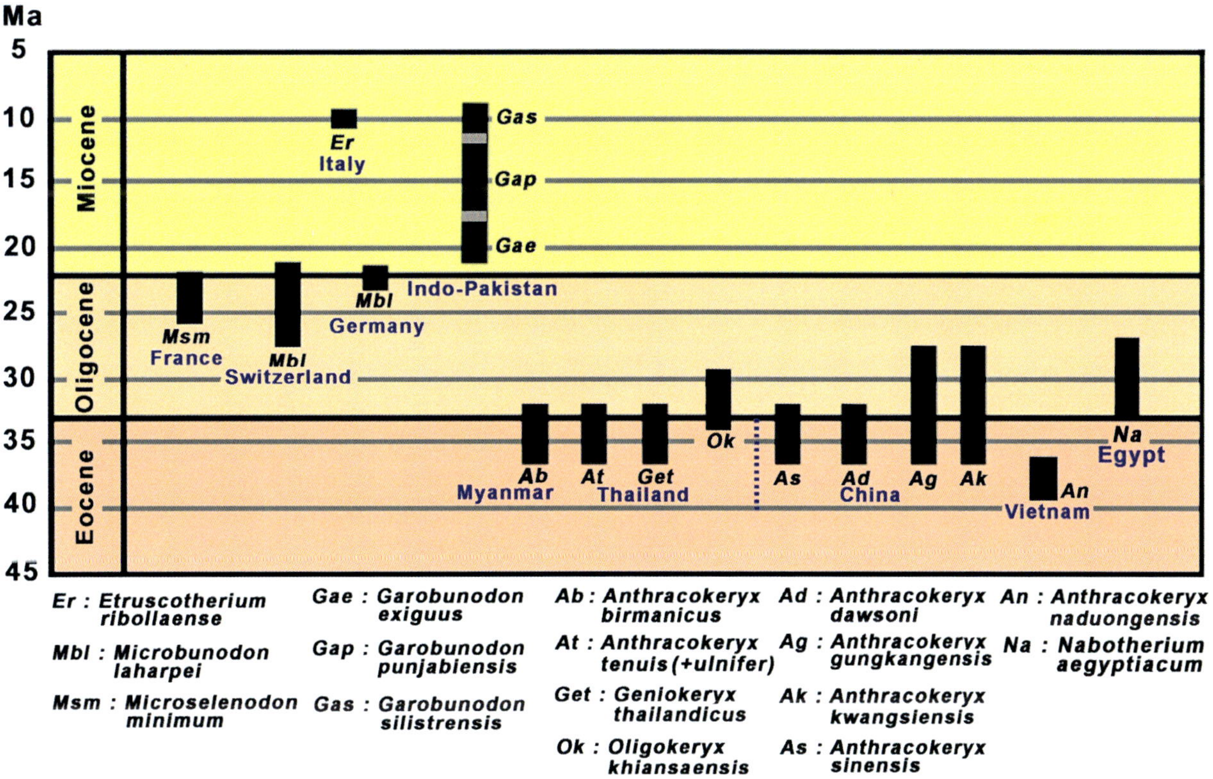

Figure 26. Chronological and geographic distribution of Microbunodontinae. Data adapted from Averianov et al. 2017; Ducrocq, 2020; Ducrocq et al. 2015, 2025; Lihoreau et al. 2004; Pickford, 2021; Qiu, 1977; Sileem et al. 2016; Tsubamoto, 2010; Wang, 1985; Zdansky, 1930. The doubtful record from Turkey (Newton, 1904) that was included in *Microbunodon* by Lihoreau et al. (2004) is removed because the specimen from Masatly, belongs to a bothriodont.

Figure 27. Occlusal views of the right D4/ in the juvenile maxilla of *Sivameryx palaeindicus* (LYDEKKER, 1877b) from Caribari, Bengal, India (NHMUK PV OR 19040) showing the disposition of the paraconule between the protocone and the paracone, and its short postparaconule crista, as well as the broad postprotocrista and endoprotocrista, the latter structure blocking off the postparaconule crista from the median transverse valley. Note also the loop-like mesostylar complex. **A,** illustration by Pentland (1828, pl. 45, fig. 3) prior to damage to the metacone and paracone; **B,** occlusal view of the specimen as it is today; **C,** nomenclature of the cusps and crests. (Mc, Metaconule; Me, Metacone; Pa, Paracone; Pc, Paraconule; Pr, Protocone). The cristae are as follows: **1,** Preprotocrista; **2,** Postprotocrista; **3,** Premetacrista; **4,** Postmetacrista; **5,** Preparaconule cristule; **6,** Postparaconule cristule; **7,** Preprotocrista; **8,** Postprotocrista; **9,** Lingual postprotocrista; **10,** Premetaconule cristule; **11,** Postmetaconule cristule; **12,** Lingual premetaconule cristule.

General discussion

As was pointed out by Forster-Cooper (1924) and subsequent authors (Matthew, 1929; Pickford, 1987) studies of the anthracotheres from Indo-Pakistan have given rise to an exceptionally complex series of interpretations, with examples of taxonomic splitting, frequent transfers of specimens between different genera and species, typographical errors, different ways of measuring teeth, confusing selection of type specimens, mis-identification of the meristic position of molars (see Annex), repetition of specific names for different taxonomic entities and inclusion of non-anthracotherian fossils in the hypodigms (Lydekker, 1885a; Forster-Cooper, 1913; Khan et al. 2013).

Much of the taxonomic confusion was due to the relative scarcity of fossils and their incomplete preservation, not just from the Indo-Pakistan region, but also of material from other parts of the world. Thus, even though authors could be confident that they were in the possession of both bunodont and relatively selenodont anthracotheres, with either four or five cusps in the upper molars, they were handicapped when endeavouring to determine the taxonomic status of the fossils. Furthermore, the initial illustration of the pentacuspidate anthracothere from Caribari (Pentland, 1828, pl. 45, fig. 3) was erroneous in the sense that it shows only four cusps in the D4/ instead of five. The condition of the specimen (NHMUK PV OR 19040) has deteriorated since it was first illustrated, but it clearly possesses a distinct paraconule with a short postparaconule cristule ('6' in Fig. 27) as was pointed out by Forster-Cooper (1924).

Pickford (1987) summarised the complex nomenclatorial history of the small bunodont anthracotheres from Indo-Pakistan, so it is not necessary to repeat it. However, more recent publications have continued to add complexity to the issues, partly due to the misreading of specimen numbers as explained above, but also by mis-reporting the discovery locus of certain specimens.

The species *Microbunodon milaensis*, erected by Lihoreau et al. (2004) is a case in point. Included in the hypodigm of the new taxon was a lower third molar (listed as NHMUK M19442) said to be from an unknown locality. However, the specimen concerned is NHMUK PV OR 19042, and it is from Caribari, the type locality of *Garobunodon silistrensis*. The tooth is close in dimensions and morphology to the m/3 in the holotype of *Microbunodon milaensis* (YPM 10999) from Dhok Mila, Pakistan, and belongs, as the authors concluded, to the same species. However, the m/3 from Caribari possesses precisely the dimensions and morphology that are compatible with the type specimen of *Garobunodon silistrensis*, which is also from Caribari. It is difficult to evade the conclusion that the two Caribari teeth belong to the same species, and that *Microbunodon milaensis* is a junior synonym of it.

Khan et al. (2013, fig. 22) attributed some fossils from the Chinji Formation to *Microbunodon silistrensis*. Of the four specimens illustrated, only one (PUPC 13/2) belongs to a small bundont anthracothere, to wit, the maxilla with D4/-M1/ (not M1/-M2/) that is compatible in dimensions with *Garobunodon silistrensis*. The tooth illustrated in their figure 22b (PUPC 13/1) consists of the rear two lophids of a left d/4 of a suid, probably *Conohyus* sp. or *Kachchhchoerus* sp. The damaged mandible (their figure 22c, PUPC 13/3) is pooorly preserved and thus difficult to identify, and the mandible (PUPC 13/4) illustrated in their figure 22d contains the d/4 and m/1 (not the m/2-m/3). The anterior tooth in this jaw is the d/4, with a characteristic accessory buccal root supporting the protoconid, and the crown broadening distally. The m/1 in this jaw is fully erupted, but being partly obscured by matrix, its morphology is difficult to discern. However, the fact that the preprotocristid extends as far as the lingual edge of the crown indicates that the specimen is more likely to belong to a species of *Hemimeryx* than to a species of *Microbunodon*. The dimensions of the teeth accord with its attribution to *Hemimeryx nanus* which is also present in the collections from the same sites discussed in the paper, but which was included in the genus *Merycopotamus* by the authors.

Conclusions

The present study revises the taxonomy and systematics of the small bundont anthracotheres from the Miocene of the Indian Subcontinent, revealing that they belong to a genus distinct from other members of the subfamily Microbunodontinae, necessitating the creation of a new genus *Garobunodon* (see Colbert, 1935, for a dicussion concerning the potential presence of a distinct genus of small bunodont anthracotheres in the subcontinent).

Small anthracotheres with bunodont, pentacuspidate, upper molars were first recorded from the Miocene deposits of Indo-Pakistan almost two centuries ago (Pentland, 1828) and they have been recovered sporadically ever since from five areas of the subcontinent (Bengal, Jammu & Kashmir, the Potwar Plateau (and Salt Range), the Bugti region and Sind). Because many of the fossils are fragmentary, the taxonomic interpretations of the specimens proved to be contentious, with some material being attributed to *Anthracotherium*, some to *Microselenodon* and some to *Microbunodon* as well as to other genera such as *Choeromeryx*, with many specimens being transferred from one genus to another by different authors. It is now clearer that all the Indo-Pakistani small, bunodont anthracothere fossils belong to a single genus, herein called *Garobunodon*, but that there are three time-succesive species present, which differ from each other in body size.

The differences in mandibular morphology that distinguish the European species of *Microbunodon* from specimens from the Indo-Pakistani attributed to this genus (Lihoreau et al. 2004), indicate that the fossils from the two regions belong to different genera, a conclusion supported by the differences in dental morphology. However, the scarcity of cranial material from Indo-Pakistan, and the rarity of the anterior parts of mandible renders comparisons difficult.

The genus *Garobunodon* possibly descended from *Geniokeryx*, known from the late Eocene and basal Oligocene of Thailand and China (Ducrocq, 2020), or possibly from *Oligokeryx*, known from the Oligocene of southern Thailand (Ducrocq et al. 2025). Dental morphology suggests that *Garobunodon* is unlikely to have descended from *Microselenodon* or *Microbunodon*.

The description of *Garobunodon* increases the diversity in the subfamily Microbunodontinae to eight genera, the previously described ones being *Microbunodon*, *Microselenodon*, *Anthracokeryx*, *Geniokeryx*, *Nabotherium*, *Oligokeryx* and *Etruscotherium*.

The oldest known species of *Garobunodon* from the Indian subcontinent is *G. exiguus*, from the Early Miocene, which was followed in time by *G. punjabiensis* from the late Early Miocene to late Middle Miocene, followed in its turn by *G. silistrensis*, which persisted from the end of the Middle Miocene into the basal phase of the Late Miocene, after which no more fossils of the genus have been found, indicating that the lineage probably went extinct in the subcontinent soon after the arrival of the equid *Hipparion* in the Old World.

Acknowledgements

Thanks to the curators for facilitating access to fossils in their care, and, in cases, for images and archival information concerning their proveninece INHMUK (Neil Adams, Spyridoula Pappa), AMNH (Michael Novacek, Carl Mehling, Chris Gilbert), Cairo (Mohamed AbdelGawad)]. Thanks also to Takehisa Tsubamoto for help in locating literature dealing with fossils from Myanmar.

References

Averianov, A, Obraztsova, E, Danilov, I. & Jin, J. 2017. Anthracotheriid artiodactyl *Anthracokeryx* and an Upper Eocene age for the Youganwo Formation of southern China.– Historical Biology, **31**: 1115 1122.

Barry, J.C. & Flynn, L.J. 1990. Key biostratigraphic events in the Siwaliks sequence. – Pp. 557-571 in: E.H. Lindsay, V. Fahlbusch, & P. Mein (Eds): European Neogene Mammal Chronology. – New York, Plenum Press.

Barry, J.C., Morgan, M.E., Flynn, L.J., Pilbeam, D., Behrensmeyer, A.K., Raza, S.M., Khan, I.A., Badgley, C., Hicks, J. & Kelley, J. 2002. Faunal and environmental change in the Late Miocene Siwaliks of northern Pakistan. – Paleobiology, **28**(**S2**): 1-71.

Bhandari, A., Pickford, M. & Tiwari, B.N. 2015. Basal Late Miocene Mammal Fauna from Tapar and Pasuda, Kachchh. – Münchner Geowissenschaftliche Abhandlungen, **43**: 1-40.

Blainville (Ducrotay de), H. 1848. Des Anthracotheriums. – Pp. 121-144 in: H. Blainville (Ed.) *Osteographies*. – Paris, J.B. Baillière et Fils.

Boettger, O. 1877. Ueber das kleine *Anthracotherium* aus der Braunkohle von Rott bei Bonn. – Palaeontographica, **24**: 163-173.

Brunet, M. 1968. Découverte d'un crâne d'Anthracotheriidae, *Microbunodon minimum* (Cuvier), à la Milloque (Lot-et-Garonne). – Comptes Rendus de l'Académie des Sciences Paris, **267**: 835-838.

Cabard, P. 1976. Monographie du genre *Microbunodon* Depéret, 1908 (Mammalia, Artiodactyla, Anthracotheriidae) de l'Oligocène supérieur d'Europe de l'Ouest. – PhD Dissertation. Université de Poitiers, 150 pp.

Colbert, E.H. 1933. An Upper Tertiary Peccary from India. – American Museum Novitates, **635**: 1-7.

Colbert, E.H. 1935. Siwalik mammals in the American Museum of Natural History. – Transactions of the American Philosophical Society, **26**: 1-401.

Cope, E.D. 1885. On the Evolution of the Vertebrata, Progressive and Retrogressive. – American Naturalist, **19** (**2**): 140-148; (**3**): 234-247; (**4**): 341-353.

Cuvier, G. 1822. *R*echerches sur les ossemens fossiles, où l'on rétablit les caractères de plusieurs animaux dont les révolutions du globe ont détruit les espèces. – Paris, E. d'Ocagne, Volume **3**, 405 pp., plates 1-80.

Cuvier, G. 1824. Recherches sur les ossemens fossiles, où l'on rétablit les caractères de plusieurs animaux dont les révolutions du globe ont détruit les espèces. – Paris, E. d'Ocagne, Volume **5**, part ii, 547 pp. plates 1-33.

Depéret, C. 1907. Les Transformations du Monde Animal. – Paris, E. Flammarion, 300 pp.

Depéret, C. 1908. L'histoire géologique et la phylogénie des Anthracotheridés. – Comptes Rendus Hebdomadaires de l'Académie des Sciences Paris, **146**: 158-162.

Ducrocq, S. 2019. *Pakkokuhyus* and *Progenitohyus* (Artiodactyla, Mammalia) from the Eocene of Southeast Asia are not Helohyidae : palaeobiogeographical implications. – Paläontologische Zeitschrift, **93** (**1**): 105-113.

Ducrocq, S. 2020. Taxonomic revision of *Anthracokeryx thailandicus* Ducrocq, 1999. (Anthracotheriidae, Microbunodontinae) from the Upper Eocene of Thailand. – Vertebrata PalAsiatica, **50** (**4**): 293-304.

Ducrocq, S., Chaimanee, Y., Yamee, C. & Jaeger, J.-J. 2025. A new anthracothere (Hippopotamoidea, Anthracotheriidae) from Oligocene deposits of southern Thailand. – Journal of Vertebrate Paleontology, **44** (**5**). https://doi.org/10.1080/02724634.2025.2467012

Ducrocq, S., Benammi, M., Chavasseau, O., Chaimanee, Y., Suraprasit, K., Pha, P.D., Le Phuong, V., Van Phach, P. & Jaeger, J.-J. 2015. New Anthracotheriidae (Cetartiodactyla, Mammalia) from the Paleogene of northeastern Vietnam: biochronological implications. – Journal of Vertebrate Paleontology, **35** (**3**): e929139. https://doi.org/10.1080/02724634.2014.929139

Falconer, H. 1868. Palaeontological Memoirs. – London, Robert Hardwicke, Volume 1 (lvi+590); Volume 2 (xiii+675).

Falconer, H. & Cautley, P.T. 1847. Fauna Antiqua Sivalensis. – 136 pp. Atlas. Plates 1-92 + A-R + A-B, London.

Filhol, H. 1889. Observations relatives à la dentition inférieure de l'*Anthracotherium minimum*. – Bulletin de la Société Philomathique de Paris, **8** (**1**): 51-53.

Filhol, H. 1891a. De la dentition supérieure de l'*Antracotherium minimum*. – Bulletin de la Société Philomathique de Paris, **3**: 89-91.

Filhol, H. 1891b. Note concernant l'étude d'une tête d'*Anthracotherium minimum* (CUVIER). – Bulletin de la Société Philomathique de Paris, **3**: 162-163.

Forster-Cooper, C. 1913. New anthracotheres and allied forms from Baluchistan. Preliminary notice. – Annals and Magazine of Natural History, **12** (**8**): 514-522.

Forster-Cooper, C. 1924. The Anthracotheriidae of the Dera Bugti deposits in Baluchistan. – Memoirs of the Geological Survey of India, Palaeontologia Indica, **8**: 72.

Gervais, 1849 (1848-1852). Zoologie et Paléontologie françaises (animaux vertébrés). – Paris, Arthus Bertrand, 544 pp. 84 plates.

Ghezzo E. & Giusberti L. 2016. New insights on *Anthracotherium monsvialense* De Zigno, 1888 (Mammalia, Cetartiodactyla) from the lower Oligocene of Monteviale (Vicenza, Northeastern Italy). – Rivista Italiana di Paleontologia e Stratigrafia, **122** (**3**): 119-140.

Goldi, E.A. 1914. Die Tierwelt der Schweiz in der Gegenwart und in der Vergangenheit. Band 1. Wirbeltiere. – Bern, Verlag von A. Francke, 654 pp.

Hellmund, M. 1991. Revision der Europäischen species der Gattung *Elomeryx* MARSH 1894 (Anthracotheriidae, Artiodactyla, Mammalia) Odontologische Untersuchungen. – Palaeontographica, Abteilung A, **220**: 1-101.

Holroyd, P.A. & Ciochon, R.L. 1995. A new artiodactyl (Mammalia) from the Eocene Pondaung Sandstones, Burma. Annals of the Carnegie Museum, **64** (**3**): 177-183.

Hünermann, K.A. 1967. Der Schädel von *Microbunodon minus* (Cuvier) (Artiodactyla, Anthracotheriidae) aus dem Chatt (Oligozän). – Eclogae geologicae Helvetiae, **60** (**2**): 661-688.

Khan, M.A., Akhtar, M., Babar, M.A., Abbas, S.G., Siddiq, M.K., Nawaz, M.K., Mubashar, M., Yaqoob, H., Nazir, S., Farheen, S., Draz, O., Shahbaz, S., Sultana, N., Zahra, N. & Noreen, S. 2013. Some new remains of Middle Miocene mammals from the Chinji Formation, Northern Pakistan. – Pakistan Journal of Zoology, Supplementary Series, **12**: 1-55.

Kowalevsky, W. 1874. Monographie der Gattung *Anthracotherium* Cuv. und Versuch einer natürlichen Classification der fossilen Hufthiere. – Palaeontographica, **22**: 131-347, tabs 7-17.

Leidy, J. 1869. The extinct mammalian fauna of Dakota and Nebraska, including an account of some allied forms from other localities, together with a synopsis of the mammalian remains of North America. – Journal of the Academy of Natural Sciences, Philadelphia, **7**: 1-472.

Lihoreau, F. 2003. Systématique et Paléoécologie des Anthracotheriidae [Artiodactyla; Suiformes] du Mio-Pliocène de l'Ancien Monde: Implications Paléobiogéographiques. – PhD Thesis, University of Poitiers, 395 pp.

Lihoreau, F. & Ducrocq, S. 2007. Family Anthracotheriidae. In: In: Prothero D.R. & Foss S.E. (Eds). The Evolution of Artiodactyls. – Baltimore, The Johns Hopkins University Press, pp. 89-105.

Lihoreau, F., Blondel, C., Barry, J. & Brunet, M. 2004. A new species of the genus *Microbunodon* (Anthracotheriidae, Artiodactyla) from the Miocene of Pakistan: genus revision, phylogenetic relationships and palaeobiogeography. Zoologica Scripta, **33** (**2**): 97-115.

Lydekker, R. 1877a. Notices of new or rare mammals from the Siwaliks. – Records of the Geological Survey of India, **10**: 76-83.

Lydekker, R. 1877b. Note on the genera *Choeromeryx* and *Rhagatherium*. – Records of the Geological Survey of India, **10**: 225.

Lydekker, R. 1878. Notices of Siwalik mammals. – Records of the Geological Survey of India, **11**: 64-104.

Lydekker, R. 1883. Siwalik selenodont Suina. – Palaeontologia Indica, **10**: 143-177.

Lydekker, R. 1885a. Catalogue of the Fossil Mammalia in the British Museum (Natural History), Part II. Containing the Order Ungulata, Suborder Artiodactyla. – London, The Trustees of the British Museum, pp. i-xxii + 1-324.

Lydekker, R. 1885b. Note on a third species of *Merycopotamus*. – Records of the Geological Survey of India, **18**: 145-146.

Matthew, W.D. 1929. Critical observations upon Siwalik mammals. – Bulletin of the American Museum of Natural History, **56**: 437-560.

Matthews, S.C. 1973. Notes on open nomenclature and on synonymy lists. – Palaeontology, **16** (**4**): 713-719.

Miller, E.R., Gunnell, G.F., Abdel Gawad, M., El-Barkooky, A.M., Clementz, M. & Hassan, S.M. 2014. Anthracotheres from Wadi Moghra, early Miocene, Egypt. – Journal of Paleontology, **88** (**5**): 967-981

Newton, R.B. 1904. Notes on the Post-Tertiary and Tertiary fossils obtained by Col. English from the district surrounding the Dardanelles. – Quarterly Journal of the Geological Society of London, **60**: 277-292.

Owen, R. 1848. Description of teeth and portions of jaw of two extinct anthracotherioid quadrupeds (*Hyopotamus vectianus* and *Hyop. bovinus*) discovered by the Marchioness of Hastings in the Eocene deposits of the N.W. coast of the Isle of Wight: with an attempt to develop Cuvier's idea of the classification of Pachyderms by the number of their toes. – Quarterly Journal of the Geological Society of London, **4**: 103-141.

Patel, B.A., Gilbert, C., Sehgal, R.K., Singh, N.P., Deepak Chaudhry, A.D., Wazir, W.A., Campisano, C.J. & Patnaik, R. 2022. A virtual repository of Lower Siwalik Miocene fossils from Ramnagar (Udhampur District, Jammu & Kashmir) India. – Abstract Book, 91st Annual Meeting of the American Association of Biological Anthropologists, pp. 141-142.

Patnaik, R. 2016. Neogene-Quaternary Mammalian Paleobiogeography of the Indian Subcontinent: an appraisal. – Comptes Rendus Palevol, **15**: 889-902.

Pentland, J.B. 1828. Description of fossil remains of some animals from the North-east border of Bengal. – Transactions of the Geological Society of London, **2**: 393-394.

Pickford, M. 1981. Preliminary Miocene Mammalian biostratigraphy for Western Kenya. – Journal of Human Evolution, **10**: 73-97.

Pickford, M. 1987. Révision des suiformes (Artiodactyla, Mammalia) de Bugti (Pakistan). – Annales de Paléontologie, **73**: 289-350.

Pickford, M. 1988. The age(s) of the Bugti fauna(s). – Pp. 937-955 in: The Palaeoenvironment of East Asia from the Mid-Tertiary. – University of Hong Kong Press.

Pickford, M. 1991. Revision of the Neogene Anthracotheriidae of Africa. In: M.J. Salem, O.S. Hammuda, & M.T. Eliagoubi (Eds) The Geology of Libya. – Elsevier, Amsterdam, Volume **4**: 1491-1525.

Pickford, M. 2016. Anthracotheres from the Oligocene of Aubenas-les-Alpes, France. – Annales de Paléontologie, **102**: 243-260.

Pickford, M. 2021. Europe's last anthracothere (Artiodactyla, Mammalia) from Ribolla (MN 12) Italy. – Historical Biology, 9 pp. https://doi.org/10.1080/08912963.2021.190 0169

Pickford, M. & Chaïd-Saoudi, Y. 2024. The biostratigraphic puzzle of Bou Hanifia (latest Miocene, Algeria, Terrestrial fauna). – Fossil Imprint, **80** (**2**): 319-328.

Pickford, M. & Rogers, D. 1987. Révision des Suiformes de Bugti, Pakistan. – Comptes Rendus de l'Académie des Sciences Paris, **305**: 634-646.

Pickford, M. & Senut, B. 2003. Miocene palaeobiology of the Orange River valley, Namibia. – Memoirs of the Geological Survey of Namibia, **19**: 1-22.

Pilgrim, G.E. 1908. The Tertiary and Post-Tertiary freshwater deposits of Baluchistan and Sind with notices of new vertebrates. – Records of the Geological Survey of India, **37**: 139-169.

Pilgrim, G.E. 1910a. Notice of new mammalian genera and species from territories of India. – Records of the Geological Survey of India, **40**: 63-71.

Pilgrim, G.E. 1910b. Preliminary note on a revised classification of the Tertiary freshwater deposits of India. – Records of the Geological Survey of India, **40**: 185-207.

Pilgrim, G.E. 1912. The vertebrate fauna of the Gaj series in the Bugti Hills and the Punjab. – Memoirs of the Geological Survey of India, Palaeontologia Indica, **4**: 1-83.

Pilgrim, G.E. 1913. The correlation of the Siwaliks with mammal horizons of Europe. – Records of the Geological Survey of India, **43**: 264-326.

Pilgrim, G.E. 1926. The fossil Suidae of India. – Memoir of the Geological Survey of India, Palaeontologia Indica, **8** (**4**): 1-68.

Pilgrim, G.E. 1928. The Artiodactyla of the Eocene of Burma. – Palaeontologia Indica, New Series, **8**: 1-39.

Pilgrim G.E. & Cotter G.P. 1916. Some newly discovered Eocene mammals from Burma. – Records of the Geological Survey of India, **47**: 42-77.

Pomel, A. 1848. Recherches sur les caractères et les rapports entre eux des divers genres vivants et fossiles des mammifères ongulés. – Comptes Rendus Hebdomadaires des Séances de l'Académie des Sciences, Paris, **24**: 686-688.

Qiu, Z. 1977. Notes on the new species of *Anthracokeryx* from Guanxi. – Vertebrata PalAsiatica, **15** (**1**): 54-58.

Raza, S.M., Barry, J.C., Meyer, G.E. & Martin, L. 1984. Preliminary report on the geology and vertebrate fauna of the Miocene Manchar Formation, Sind, Pakistan. – Journal of Vertebrate Paleontology, **4**: 584-599.

Raza, S.M. & Meyer, G.E. 1984. Early Miocene geology and paleontology of the Bugti Hills. – Memoir of the Geological Survey of Pakistan, **11**: 43-63.

Renevier, E. 1879. Les *Anthracotherium* de Rochette. – Bulletin de la Société vaudois des sciences naturelle, **16** (**2**): 140-148.

Rutimeyer, 1857. Über *Anthracotherium magnum* und *hippoideum*. – Neue Denkschriften der Allgemeinen Schweizerischen Gesellschaft für die Gesammten Naturwissenschaften, Zurich. **15** (**8**): 1-32.

Sarigül, V. 2021. A Short History of Paleontology in Turkey, Part I: From the Nineteenth Century to the collapse of Ottoman Turkey. – Earth Sciences History, **40** (**1**): 158-201.

Sickenberg, O. 1934, Die ersten Reste von Landsäugetieren aus den Linzer Sanden. – Verhandlungen der Geologischen Bundesanstalt, **1934**: 60-63.

Sileem, A.H., Sallam, H.M., Hewaidy, A.G.A., Miller, E.R. & Gunnell, G.F. 2016. A new anthracothere (Artiodactyla) from the early Oligocene, Fayum, Egypt, and the mystery of African '*Rhagatherium*' solved. – Journal of Paleontology, **90**: 170-181.

Stehlin, H.G. 1910. Zur revision der europäischen Anthracotherien. – Verhandlungen der Naturforschenden Gesellschaft, **21**: 165-185.

Steininger, F.F. 1999. Chronostratigraphy, Geochronology and Biochronology of the Miocene European Land Mammal Mega-Zones (ELMMZ) and the Miocene Mammal Zones (MN-Zones). – Pp. 9-24 in: Rössner, G.E. & Heissig, K. (Eds). The Miocene Land Mammals of Europe.– München, Pfeil-Verlag.

Teller, F. 1884. Neue Anthracotherienreste aus Südsteiermark und Dalmatien. – Beiträge zur Paläontologie Österreich-Ungarns und des Orients, Wien, **4**: 45-134.

Troschel, F.H. 1859. Untitled. Verhandlungen des Naturhistorischen Vereins der Preussischen Rheinlande und Westphalens, **16**. – Correspondenzblatt N° 3 Bericht über die 16 General-Versammlung zu Bonn, p. 49.

Trouessart, E. 1904. Catalogus mammalium tam viventium quam fossilium Quinquennale supplementum. – Paris, R. Friedlander & Sohn, 929 pp.

Tsubamoto, T. 2010. Recognition of *Microbunodon* (Artiodactyla, Anthracotheriidae) from the Eocene of China. – Paleontological Research, **14** (**2**): 161-165.

Tsubamoto, T., Takai, M., Egi, N., Shigehara, N., Tun, S.T., Aung, A.K. Soe, A.N. & Thein, T. 2002. The Anthracotheriidae (Mammalia; Artiodactyla) from the Eocene Pondaung Formation (Myanmar) and comments on some other anthracotheres from the Eocene of Asia. – Paleontological Research, **6** (**4**): 363-384.

Vasishat, R.N., Gaur, R. & Chopra, R.K. 1978. Community structure of middle Siwalik vertebrates from Haritalyangar (H.P.) India. – Palaeogeography, Palaeoclimatology, Palaeoecology, **23**: 131-140.

Von Meyer, H. 1852. Untitled. – Neues Jahrbuch für Mineralogie, Geognosie, Geologie und Petrefaktenkunde, Stuttgart, **1852**: 300-306.

Wang, J. 1985. A new species of *Anthracokeryx* from Shanxi. – Vertebrata PalAsiatica, **23** (**1**): 53-59 (in Chinese with English Summary).

Xu, Y-X. 1962. Some new anthracotheres from Shansi and Yunnan. – Vertebrata PalAsiatica **6** (**3**): 232-245.

Zdansky, O. 1930. Die alttertiären Säugetiere Chinas nebst stratigraphischen Bemerkungen. – Palaeontologia Sinica, **6** (**2**): 5-87.

Zong, G., Chen, W., Huang, X. & Xu, Q. 1996. Cenozoic Mammals and Environment of Hengduan Mountains Region. – Beijing, China Ocean Press, 279 pp. (in Chinese with English Abstract).

ANNEX

Measurements (in mm) of the teeth of *Garobunodon* from Indo-Pakistan

BLB, bucco-lingual breadth; **MDL**, mesio-distal length; **est.**, estimated. (**Own**: the author's measurements: *A. Anthracotherium*; *G. Garobunodon*, *M. Microbunodon*). Continued on next pages.

Catalogue N°	Tooth	MDL	BLB	Locality	Taxon this paper	Data sources and comments
NHMUK M 12704	m/2 lt	12.6	8.3	Chur Lando	*G. exiguus*	Own, 2025; Lihoreau et al. 2004 as 12.3x8, *M. silistrensis*; Pickford, 1987;
YPM 21078	m/2 lt	12.6	8.9	Chinji	*G. exiguus*	Lihoreau et al. 2004 as *M. silistrensis*
NHMUK M 12705	m/2 rt	13.5	9.3	Chur Lando	*G. exiguus*	Pickford, 1987 as 13x9; Lihoreau et al. 2004 as 13x9.2, *M. silistrensis*
NHMUK M 12706	m/2 rt	12.4	8.9	Chur Lando	*G. exiguus*	Own, 2025; Lihoreau et al. 2004 as 13x8.9, as *M. silistrensis*; Pickford, 1987;
NHMUK M 12704	m/3 rt	18.3	9.1	Chur Lando	*G. exiguus*	Own, 2025; Lihoreau et al. 2004, as 18.8x9.3, as *M. silistrensis*
NHMUK M 12705	m/3 rt	20.0	9.4	Chur Lando	*G. exiguus*	Own, 2025; Lihoreau et al. 2004 as *M. silistrensis*; Pickford, 1987
NHMUK M 12706	m/3 rt	20.4	10.7	Chur Lando	*G. exiguus*	Own, 2025; Lihoreau et al. 2004 as 20.4x9.9, as *M. silistrensis*; Pickford, 1987;
NHMUK M 12054	M1/ lt	10.0	10.0	Dera Bugti	*G. exiguus*	Own, 2025 est. from fig. Forster-Cooper, 1924, fig. 18c
NHMUK M 12054	M2/ lt	12.4	11.5	Dera Bugti	*G. exiguus*	Own, 2025 est. from fig. Forster-Cooper, 1924, fig. 18c
NHMUK M 12055	M2/ rt	11.3	11.9	Dera Bugti	*G. exiguus*	Own, 2025 est. from fig. Forster-Cooper, 1924, fig. 18b
NHMUK M 12053	M3/ rt	–	–	Dera Bugti	*G. exiguus*	Forster-Cooper, 1924, fig. 18a
NHMUK M 12055	M3/ rt	14.4	14.1	Dera Bugti	*G. exiguus*	Own, 2025, est. from fig. in Forster-Cooper, 1924
PUPC 13/2	D4/ lt	9.5	9.0	Chinji	*G. punjabiensis*	Khan et al. 2013, fig. 22a
YPM 45725	m/1 lt	10.0	7.0	Chinji	*G. punjabiensis*	Lihoreau et al. 2004 as *M. silistrensis*
GSI B 466 (NHMUK M 11059)	m/1 rt	10.5	7.4	Khumbi	*G. punjabiensis*	Own, 2025; Pilgrim 1912, pl. 14, fig. 3 as 10x7, *Microselenodon mus* holotype; Lihoreau et al. 2004 as 9,6x7,2 as *M. silistrensis*
YPM 28194	m/1 rt	11.6	–	Chinji	*G. punjabiensis*	Lihoreau et al. 2004 as *M. silistrensis*
NHMUK M 12708	m/2 lt	14.5	9.1	Chur Lando	*G. punjabiensis*	Lihoreau et al. 2004 as 12.9x9, as *M. silistrensis*; Pickford, 1987; Forster-Cooper, 1924, as 12.7x 9.1, fig. 16 as *A. silistrense*
YPM 30507	m/2 lt	13.5	8.3	Chinji	*G. punjabiensis*	Lihoreau et al. 2004 as m/1 in table, *M. silistrensis*
GSI B 104	m/2 rt	15.2	10.6	Hasnot?	*G. punjabiensis*	Lihoreau et al. 2004 as Siwalik Hills, *M. silistrensis*; Lydekker, 1877a holotype of *punjabiense*, Hasnot; Pilgrim 1912 pl. 14 fig. 2 as *silistrense*
GSI B 466 (NHMUK M 11059)	m/2 rt	13.6	9.2	Khumbi	*G. punjabiensis*	Own, 2025; Pilgrim 1912, pl. 14, fig. 3 as 13.5x8.7, *Microselenodon mus* holotype; Pickford, 1987; Lihoreau et al. 2004 as 12.8x9.1, *M. silistrensis*
YPM 28194	m/2 rt	14.4	10.3	Chinji	*G. punjabiensis*	Lihoreau et al. 2004 as *M. silistrensis*
YPM 42274	m/2 rt	14.3	8.0	Chinji	*G. punjabiensis*	Lihoreau et al. 2004 as *M. silistrensis*
GSP S 15	m/3 lt	22.0	10.0	Sind	*G. punjabiensis*	Lihoreau et al. 2004 as *M. silistrensis*
NHMUK M 12708	m/3 lt	21.4	10.5	Dera Bugti	*G. punjabiensis*	Own, 2025; Lihoreau et al. 2004 as 21x10.5, as Chur Lando, *M. silistrensis*; Pickford, 1987; Forster-Cooper, 1924, 20.0x10.2, fig. 16, as *A. silistrense*
YPM 21063	m/3 lt	22.6	11.2	Chinji	*G. punjabiensis*	Lihoreau et al. 2004 as *M. silistrensis*
YPM 21078	m/3 lt	21.3	10.0	Chinji	*G. punjabiensis*	Lihoreau et al. 2004 as *M. silistrensis*
YPM 26382	m/3 lt	21.0	12.0	Chinji	*G. punjabiensis*	Lihoreau et al. 2004 as *M. silistrensis*
YPM 41301	m/3 lt	20.7	10.5	Chinji	*G. punjabiensis*	Lihoreau et al. 2004 as *M. silistrensis*
GSP S 329	m/3 lt	–	10.6	Sind	*G. punjabiensis*	Lihoreau et al. 2004 as *M. silistrensis*
GSI B 104	m/3 rt	23.4	11.2	Hasnot?	*G. punjabiensis*	Lihoreau et al. 2004 as Siwalik Hills, *M. silistrensis*; Lydekker, 1878 holotype of *punjabiense*; Pilgrim, 1912 pl. 14, fig. 2 as silistrense
GSI B 466 (NHMUK M 11059)	m/3 rt	22.2	10.6	Khumbi	*G. punjabiensis*	Own, 2025; Pilgrim 1912, pl. 14, fig. 3 as *Microselenodon mus*, holotype as 22x12; Lihoreau et al. 2004 as 21.3x10.8, as *M. silistrensis*
GSP S 80	m/3 rt	–	10.1	Sind	*G. punjabiensis*	Lihoreau et al. 2004 as *M. silistrensis*
YPM 13294	m/3 rt	22.5	10.8	Chinji	*G. punjabiensis*	Lihoreau et al. 2004 as *M. milaensis*
YPM 28194	m/3 rt	21.7	11.8	Chinji	*G. punjabiensis*	Lihoreau et al. 2004 as *M. silistrensis*
YPM 31564	m/3 rt	22.0	11.2	Chinji	*G. punjabiensis*	Lihoreau et al. 2004 as *M. silistrensis*
YPM 42274	m/3 rt	21.6	8.5	Chinji	*G. punjabiensis*	Lihoreau et al. 2004 as *M. silistrensis*, exceptionally narrow

Catalogue N°	Tooth	MDL	BLB	Locality	Taxon this paper	Data sources and comments
PUPC 13/2	M- / lt	11.0	11.0	Chinji	*G. punjabiensis*	Khan et al. 2013, fig. 22a
YPM 24297	M2/ lt	13.8	15.4	Chinji	*G. punjabiensis*	Lihoreau et al. 2004 as *M. silistrensis*
YPM 24297	M3/ lt	14.4	16.6	Chinji	*G. punjabiensis*	Lihoreau et al. 2004 as *M. silistrensis*
AMNH 94620	M3/ rt	15.4	15.7	Kanatti (18)	*G. punjabiensis*	Lihoreau et al. 2004 as *M. silistrensis* as 'Siwalik Hills'
GSP S 375	M3/ rt	15.0	16.0	Sind	*G. punjabiensis*	Lihoreau et al. 2004 as *M. silistrensis*
YPM 28194	p/1 rt	7.3	4.3	Chinji	*G. punjabiensis*	Lihoreau et al. 2004 as *M. silistrensis*
YPM 31838	p/1 rt	7.4	4.4	Chinji	*G. punjabiensis*	Lihoreau et al. 2004 as *M. silistrensis*
YPM 28194	p/2 rt	14.8	5.3	Chinji	*G. punjabiensis*	Lihoreau et al. 2004 as *M. silistrensis*
YPM 7421	p/2 rt	12.5	6.2	Chinji	*G. punjabiensis*	Lihoreau et al. 2004 as *M. milaensis*
YPM 28194	p/3 rt	15.0	–	Chinji	*G. punjabiensis*	Lihoreau et al. 2004 as *M. silistrensis*
YPM 45725	p/4 lt	13.0	7.8	Chinji	*G. punjabiensis*	Lihoreau et al. 2004 as *M. silistrensis*
YPM 28194	p/4 rt	13.4	7.7	Chinji	*G. punjabiensis*	Lihoreau et al. 2004 as *M. silistrensis*
YPM 47161	p/4 rt	14.0	8.0	Chinji	*G. punjabiensis*	Lihoreau et al. 2004 as *M. silistrensis*
YPM 27895	P4/ rt	10.6	11.0	Chinji	*G. punjabiensis*	Lihoreau et al. 2004 as *M. silistrensis*
WIHG RF 7	M1/ rt	12.4	12.0	Ramnagar	*G. silistrensis*	Own, 2025
AMNH FM 19444	m/1 lt	14.0	10.0	Chinji	*G. silistrensis*	Colbert, 1935 as *A. punjabiense*; Lihoreau et al. 2004 as 14.9 x 11.1
YPM 16612	m/1 lt	15.2	9.8	Nagri	*G. silistrensis*	Lihoreau et al. 2004 as *M. milaensis*
AMNH FM 94621	m/1 rt	12.8	8.9	Chinji (C44)	*G. silistrensis*	Lihoreau et al. 2004 as 'Siwalik Hills'
GSP S 377	m/1 rt	14.1	9.2	Sind	*G. silistrensis*	Lihoreau et al. 2004
WIHG FR 34/16	m/1 rt	13.6	9.2	Ramnagar	*G. silistrensis*	Own, 2025
GSP S 174	m/2	16.6	12.0	Sind	*G. silistrensis*	Lihoreau et al. 2004
AMNH FM 19444	m/2 lt	17.0	12.5	Chinji	*G. silistrensis*	Colbert, 1935 as *A. punjabiense*; Lihoreau et al. 2004 as 16.4 x 12.4
WIHG FR 33/14	m/2 lt	19.0	11.1	Ramnagar	*G. silistrensis*	Own, 2025
YPM 10999	m/2 lt	15.8	11.6	Nagri	*G. silistrensis*	Lihoreau et al. 2004, holotype of *M. milaensis*
YPM 11686	m/2 lt	18.0	12.6	Nagri	*G. silistrensis*	Lihoreau et al. 2004 as 11886 in table, *M. milaensis*
YPM 12448	m/2 lt	18.3	11.0	Nagri	*G. silistrensis*	Lihoreau et al. 2004 as *M. milaensis*
YPM 5174	m/2 lt	17.0	11.8	Nagri	*G. silistrensis*	Lihoreau et al. 2004 as *M. milaensis*
AMNH FM 94621	m/2 rt	16.0	11.0	Chinji (C44)	*G. silistrensis*	Lihoreau et al. 2004
GSP S 84	m/2 rt	16.4	11.5	Sind	*G. silistrensis*	Lihoreau et al. 2004
AMNH FM 19444	m/3 lt	24.5	14.0	Chinji	*G. silistrensis*	Colbert, 1935 as *A. punjabiense*, Lihoreau et al. 2004 as 23.4 x 13.6
NHMUK M 12707	m/3 lt	25.4	13.0	Chur Lando	*G. silistrensis*	Lihoreau et al. 2004, as *M. milaensis* unknown locality; Pickford, 1987; Forster-Cooper, 1924, fig. 17 as *punjabiense*
WIHG FR 17/1	m/3 lt	23.5	12.6	Ramnagar	*G. silistrensis*	Own, 2025
YPM 10999	m/3 lt	25.7	12.8	Nagri	*G. silistrensis*	Lihoreau et al. 2004, holotype of *M. milaensis*
YPM 12448	m/3 lt	26.7	11.3	Nagri	*G. silistrensis*	Lihoreau et al. 2004 as *M. milaensis*
NHMUK OR 19042	m/3 rt	27.0	13.4	Caribari	*G. silistrensis*	Own, 2025; Lydekker, 1885 as *Anthracotherium* sp. c, length given as 0.027; Lihoreau et al. 2004 as M19442, indet. locality 27 x 13.5, *M. milaensis*
NHMUK M 12710	M1/ rt	11.3	13.0	Chur Lando	*G. silistrensis*	Own, 2025; Lihoreau et al. 2004 as 11.5 x 13; Forster-Cooper, 1924, fig. 15c
YPM 30698	M1/ rt	12.5	13.0	Chinji	*G. silistrensis*	Lihoreau et al. 2004
NHMUK M 12709	M2/ lt	16.5	17.0	Chinji	*G. silistrensis*	Lihoreau et al. 2004 as M3/, as Bugti Hills; Pickford, 1987 as 16 x 18; Forster-Cooper, 1924, fig. 15b as Salt Range
AMNH FM 94625	M2/ rt	14.9	15.2	Ramnagar	*G. silistrensis*	Own, 2025; Lihoreau et al. 2004 as 15.5 x 15.7, as M3/

Catalogue N°	Tooth	MDL	BLB	Locality	Taxon this paper	Data sources and comments
GSI B 103	M2/ rt	17.6	18.0	Sind	*G. silistrensis*	Lydekker 1883, pl. 23, fig. 12 (est. from fig.)
NHMUK M 12710	M2/ rt	15.4	16.5	Chur Lando	*G. silistrensis*	Own, 2025; Lihoreau et al. 2004 as 15x15.7; Forster-Cooper, 1924, fig. 15c
WIHG R 23/1	M3/ lt	16.2	16.7	Ramnagar	*G. silistrensis*	Own, 2025
WIHG FR 11/1	M3/ lt	16.2	17.8	Ramnagar	*G. silistrensis*	Own, 2025
GSI B 103	M3/ rt	17.5	18.0	Sind	*G. silistrensis*	Lydekker 1883, pl. 23, fig. 10 (est. from fig.)
NHMUK M 12710	M3/ rt	15.8	18.2	Chur Lando	*G. silistrensis*	Own, 2025; Lihoreau et al. 2004 as 15.9x18; Forster-Cooper, 1924, as 16x18, fig. 15c
NHMUK OR 19041	M3/ rt	16.2	19.0	Caribari	*G. silistrensis*	Own, 2025; Lihoreau et al. 2004 as 16.1x18.8; Forster-Cooper, 1924, fig. 15a, type *A. silistrense*
WIHG FR 21/L	M3/ rt	16.4	17.6	Ramnagar	*G. silistrensis*	Own, 2025
YPM 20098	M3/ rt	16.9	18.1	Ramnagar	*G. silistrensis*	Lihoreau et al. 2004 as 15.4x16.7, *M. milaensis* listed as locality unknown and as Chinji
AMNH FM 19444	p/1 lt	8.0	4.0	Chinji	*G. silistrensis*	Colbert, 1935 as *A. punjabiense*; Lihoreau et al. 2004 as 8.8x4.2
YPM 27697	p/1 lt	12.0	4.7	Chinji	*G. silistrensis*	Lihoreau et al. 2004
YPM 10999	p/2 lt	14.4	7.0	Nagri	*G. silistrensis*	Lihoreau et al. 2004, holotype of *M. milaensis*
YPM 9795	p/2 lt	13.0	7.0	Nagri	*G. silistrensis*	Lihoreau et al. 2004 as *M. milaensis*
AMNH FM 19444	p/3 lt	15.2	8.0	Chinji	*G. silistrensis*	Colbert, 1935 as *A. punjabiense*; Lihoreau et al. 2004 as 15.2x8
YPM 10999	p/3 lt	15.0	8.0	Nagri	*G. silistrensis*	Lihoreau et al. 2004, holotype of *M. milaensis*
AMNH FM 19444	p/4 lt	15.2	9.2	Chinji	*G. silistrensis*	Colbert, 1935 as *A. punjabiense*; Lihoreau et al. 2004 as 15.2x9.2
YPM 11797	p/4 lt	15.7	9.0	Nagri	*G. silistrensis*	Lihoreau et al. 2004 as *M. milaensis*
YPM 19831	P3/ rt	15.0	9.5	Chinji	*G. silistrensis*	Lihoreau et al. 2004 as *M. milaensis*
YPM 30698	P3/ rt	13.4	8.6	Chinji	*G. silistrensis*	Lihoreau et al. 2004
YPM 9799	P3/ rt	17.4	10.0	Nagri	*G. silistrensis*	Lihoreau et al. 2004 as *M. milaensis*
YPM 19831	P4/ rt	14.0	11.0	Chinji	*G. silistrensis*	Lihoreau et al. 2004 as *M. milaensis*
YPM 30698	P4/ rt	11.3	9.0	Chinji	*G. silistrensis*	Lihoreau et al. 2004